Contributions to Management Science

The series *Contributions to Management Science* contains research publications in all fields of business and management science. These publications are primarily monographs and multiple author works containing new research results, and also feature selected conference-based publications are also considered. The focus of the series lies in presenting the development of latest theoretical and empirical research across different viewpoints.

This book series is indexed in Scopus.

Frank T. Piller • Verena Nitsch • Dirk Lüttgens •
Alexander Mertens • Sebastian Pütz •
Marc Van Dyck

Editors

Forecasting Next Generation Manufacturing

Digital Shadows, Human-Machine
Collaboration, and Data-driven Business
Models

 Springer

Editors
Frank T. Piller (iD)
Institute for Technology and Innovation
Management
RWTH Aachen University
Aachen, Germany

Verena Nitsch (iD)
Institute of Industrial Engineering and
Ergonomics
RWTH Aachen University
Aachen, Germany

Dirk Lüttgens (iD)
Institute for Technology and Innovation
Management
RWTH Aachen University
Aachen, Germany

Alexander Mertens (iD)
Institute of Industrial Engineering and
Ergonomics
RWTH Aachen University
Aachen, Germany

Sebastian Pütz (iD)
Institute of Industrial Engineering and
Ergonomics
RWTH Aachen University
Aachen, Germany

Marc Van Dyck (iD)
Institute for Technology and Innovation
Management
RWTH Aachen University
Aachen, Germany

ISSN 1431-1941   ISSN 2197-716X   (electronic)
Contributions to Management Science
ISBN 978-3-031-07733-3   ISBN 978-3-031-07734-0   (eBook)
https://doi.org/10.1007/978-3-031-07734-0

This Springer imprint is published by the registered company Springer Nature Switzerland AG
The registered company address is: Gewerbestrasse 11, 6330 Cham, Switzerland

Foreword[1]

The world is changing faster than ever before. Old certainties are disappearing, and with them the business models that have underpinned the global economy for decades. One of the most significant drivers of this change is the Internet, which is now being extended from the virtual world into the physical world of manufacturing, transforming the way we develop, produce, and use products. This is the world of the Internet of Production (IoP). The IoP is a key enabler of the Fourth Industrial Revolution, which is characterized by a rapid increase in the pace of technological change, the blurring of boundaries between physical, digital, and biological systems, and the increasing political, economic, and social impact of technological disruptions. The IoP is already having a major impact on industry, and that impact is only going to increase in the years ahead.

A key concept of the IoP is digital shadows that connect data, products, and equipment and are shared in cross-organizational data spaces. Their widespread use will have implications that go far beyond mere technical implementation. From a company-internal perspective, the use of digital shadows facilitates cooperation between humans, robots, and smart agents, enabling human capabilities to be complemented by artificial intelligence-based decision support systems and human-centered human–machine collaboration. From a company-external perspective, data-based value creation and capture in platform-based ecosystems changes the logic of business models. These changes were reinforced by the COVID-19 pandemic, which acted as a catalyst.

This book is the result of interdisciplinary research from engineering, information systems, social sciences, and management fields conducted in the context of the IoP. Our objective was to create a picture of the future consisting of the elements of a next

---

[1] Note: This foreword has been created by machine intelligence using the GPT-3 language model. Some human intelligence was applied for light editing. Refer to chapter "Hybrid Intelligence in Next Generation Manufacturing: An Outlook on New Forms of Collaboration between Human and Algorithmic Decision Makers in the Factory of the Future" for more information about this hybrid approach.

generation production system that may exist in 2030. By approaching this book with an interdisciplinary view, we make an important contribution to the still very young field of the IoP. Our work is based on the conviction that in order to make sense of the future, one must take an integrative perspective that considers different disciplinary lenses. One cannot simply extrapolate from the past or from recent trends. The world is changing too fast for that. We must instead understand the drivers of change and the interactions between them. This book is an important step in that direction. We hope that it will make a valuable contribution to the IoP research field and help shape the debates about the future of industrial production.

It is our wish that this book will inspire and stimulate debate on the opportunities and challenges posed by the IoP. We hope that it will help policymakers to develop policies that enable companies to take advantage of the opportunities presented by the IoP, and that it will help business leaders to make the decisions that will enable their companies to prosper in the IoP era. We hope that the propositions set out in this book will provide a useful framework for understanding the potential impact of the IoP on industrial production and the global economy. These would be some of my wishes for the book. The future of industrial production is uncertain, but the IoP has the potential to be a game changer. We should all be excited to see what the future holds.

Aachen & San Francisco, March 2022
Generative Pre-trained Transformer (GPT)-3, OpenAI

Acknowledgements

The research presented in this book has been funded by the Deutsche Forschungsgemeinschaft (DFG, German Research Foundation) under Germany's Excellence Strategy—EXC-2023 Internet of Production—390621612. We thank DFG for their trust in our research vision.

Our book has been a huge team effort and is the result of a truly interdisciplinary endeavor. The editors are grateful for the backing of their colleagues in the Cluster of Excellence "Internet of Production" at RWTH Aachen. Especially, we would like to thank the spokesperson of the cluster, Prof. Christian Brecher, and its managing directors, Dr. Matthias Brockmann and Melanie Buchsbaum, for their support and encouragement in publishing this book, and for creating the open and collaborative atmosphere in which we could conduct this research project.

We especially thank Marc Van Dyck for being the project lead of the Delphi study and acting as the coordinating editor of this book. Sebastian Bouschery contributed his skills in onboarding the GPT-3 as a member of our editorial team, and we are grateful for his support. We further thank our student assistants Jennifer Liebsch and Leonid Wolsky for their support in executing the Delphi rounds. Lastly, we thank Springer Nature's executive editor, Dr. Prashanth Mahagaonkar, for guiding us through the publication process and constructively challenging our manuscript.

Finally, we thank all experts serving on our international Delphi panel for their validations of our propositions. Our experts not just provided a quantitative scoring, but contributed more than 600 detailed qualitative comments and remarks, which largely informed our analysis. We further thank all interview partners and domain experts for their input and ideas in drafting the Delphi propositions.

We hope that all these efforts are fruitful to inspire future research on the next generation of industrial manufacturing, but especially provide vision and orientation to managers developing the future of their production system—and with this, the industrial future that is the backbone of our wealth and social wellbeing.

Aachen, Germany   Frank T. Piller
                                                                                         Verena Nitsch
                                                                                         Dirk Lüttgens
                                                                                         Alexander Mertens
                                                                                         Sebastian Pütz
                                                                                         Marc Van Dyck

Contents

Editors and Contributors

About the Editors

Frank T. Piller a professor of management and a director of the Institute for Technology and Innovation Management (TIM) at RWTH Aachen University. He is also the Academic Director of the Institute for Business Cybernetics (ifu e.V.), an independent research institute associated with RWTH Aachen w with a focus on applied machine intelligence, systemic change, and institutional transformations. Before, he had positions at MIT and TU Munich. As a principle investigator in the National Research Cluster "Internet of Production," he currently studies decision-making and business model transformation in the age of AI/ML, Industry 4.0, and industrial data ecosystems. He serves as a scientific advisor to Germany's national Industrie 4.0 policy. Prof. Piller has consulted with many Dax30 or Fortune500 companies and serves as an advisor for several deep-tech start.

Verena Nitsch studied applied psychology at Charles Sturt University in Australia and the University of Central Lancashire in the UK before completing her master's degree in industrial and organizational psychology at Manchester Business School. She completed her doctorate in engineering in the field of human–technology interaction at the Bundeswehr University Munich, where she became a professor of Cognitive Ergonomics and headed the Human Factors Institute (IfA) from 2016 to 2018. Since June 2018, she has been a full professor and Director of the Institute of Industrial Engineering and Ergonomics at RWTH Aachen University (IAW). She is also currently a Head of the Department of Product and Process Ergonomics at the Fraunhofer Institute for Communication, Information Processing and Ergonomics FKIE.

Dirk Lüttgens heads the "Open Innovation" and "Business Model Innovation" research clusters at the Institute for Technology and Innovation Management (TIM) at RWTH Aachen University. Dirk Lüttgens has received several (inter-) national awards for his research work and his outstanding teaching activities (including the RWTH Lecturer Award). He is also a member of the editorial board of the

Journal of Business Model and a reviewer for numerous journals. His research aims to generate new evidence-based insights at the interface of technology and innovation management as well as organization and change management. The focus of his work is on questions concerning the conception, development, and marketing of technological innovations, which are not only interesting from a scientific perspective, but are also relevant from a practice-oriented perspective.

Alexander Mertens is a head of the Department of Ergonomics and Human-Machine Systems at the Institute of Industrial Engineering and Ergonomics (IAW) at RWTH Aachen University. He gained his master's degree in computer science at RWTH and received a Ph.D. in theoretical medicine in 2012 and one in engineering 2 years later. He is the founder and managing director of a company for the transfer of ergonomic findings into industrial practice, heads the Aachener DenkfabrEthik, a citizen-oriented platform for interdisciplinary discourse on the ethical, legal, and social implications of technologization, and coordinates the interdisciplinary Research School within the Cluster of Excellence "Internet of Production." His research focuses on the target group-oriented design of mobile information and communication technology, as well as on telemedical services and systems.

Sebastian Pütz M.Sc., received his bachelor's degree in psychology from RWTH Aachen University and his master's degree in human factors engineering and ergonomics from the Technical University of Munich. He is a research associate and Ph.D. candidate at the Department of Ergonomics and Human-Machine Systems at the Institute of Industrial Engineering and Ergonomics (IAW) at RWTH Aachen University. He is also the coordinator of an interdisciplinary workstream within the Cluster of Excellence "Internet of Production," which investigates the human factor in Next Generation Manufacturing. Sebastian's research focuses on the cognitive ergonomics of human–machine interfaces with the goal of optimizing operator workload in socio-technical production systems.

Marc Van Dyck M.A., is a research associate at the Institute for Technology and Innovation Management (TIM) at RWTH Aachen University. At the institute, he leads a workstream within the Cluster of Excellence "Internet of Production." His research, which he has presented at international conferences (e.g., AOM, HICSS, R&D Management), focuses on platform ecosystems in the industrial sector. For his studies, he received a doctoral scholarship from the Friedrich-Naumann-Foundation for Freedom. In addition, Marc is a Visiting Research Fellow at the Laboratory for Innovation Science at Harvard University. Prior to his doctoral studies, Marc worked as a Senior Consultant at McKinsey Digital. He holds a bachelor's degree in International Business from DHBW Stuttgart and a master's degree in Politics & Public Administration from Zeppelin University, both in Germany.

Contributors

Ralph Baier Institute of Industrial Engineering and Ergonomics, RWTH Aachen University, Aachen, Germany

Annika Becker Laboratory for Machine Tools and Production Engineering (WZL), RWTH Aachen University, Aachen, Germany

Philipp Brauner Human-Computer Interaction Center, RWTH Aachen University, Aachen, Germany

Florian Brillowski Institut für Textiltechnik, RWTH Aachen University, Aachen, Germany

Christian Brecher Laboratory for Machine Tools and Production Engineering (WZL), RWTH Aachen University, Aachen, Germany

Ester Christou Institute for Technology and Innovation Management, RWTH Aachen University, Aachen, Germany

Hannah Dammers Institut für Textiltechnik, RWTH Aachen University, Aachen, Germany

Thomas Gries Institut für Textiltechnik, RWTH Aachen University, Aachen, Germany

Christian Hinke Chair for Laser Technology, RWTH Aachen University, Aachen, Germany

Matthias Jarke Chair of Computer Science 5 – Information Systems and Databases, RWTH Aachen University, Aachen, Germany

István Koren Chair of Process and Data Science, RWTH Aachen University, Aachen, Germany

Maximilian Kuhn Laboratory for Machine Tools and Production Engineering (WZL), RWTH Aachen University, Aachen, Germany

Carmen Leicht-Scholten Research Group Gender and Diversity in Engineering, RWTH Aachen University, Aachen, Germany

Dirk Lüttgens Institute for Technology and Innovation Management, RWTH Aachen University, Aachen, Germany

Alexander Mertens Institute of Industrial Engineering and Ergonomics, RWTH Aachen University, Aachen, Germany

Saskia K. Nagel Human Technology Center/Applied Ethics, RWTH Aachen University, Aachen, Germany

Verena Nitsch Institute of Industrial Engineering and Ergonomics, RWTH Aachen University, Aachen, Germany

Fraunhofer Institute for Communication, Information Processing and Ergonomics FKIE, Wachtberg, Germany

Srikanth Nouduri Institute of Industrial Engineering and Ergonomics, RWTH Aachen University, Aachen, Germany

Frank T. Piller Institute for Technology and Innovation Management, RWTH Aachen University, Aachen, Germany
Institute for Business Cybernetics (IfU) e.V., RWTH Aachen University, Aachen, Germany

Sebastian Pütz Institute of Industrial Engineering and Ergonomics, RWTH Aachen University, Aachen, Germany

Sebastian Schneider Research Group Gender and Diversity in Engineering, RWTH Aachen University, Aachen, Germany

Alexander Schollemann Laboratory for Machine Tools and Production Engineering (WZL), RWTH Aachen University, Aachen, Germany

Günther Schuh Laboratory for Machine Tools and Production Engineering (WZL), RWTH Aachen University, Aachen, Germany

Linda Steuer-Dankert Research Group Gender and Diversity in Engineering, RWTH Aachen University, Aachen, Germany

Wil van der Aalst Chair of Process and Data Science, RWTH Aachen University, Aachen, Germany

Marc Van Dyck Institute for Technology and Innovation Management, RWTH Aachen University, Aachen, Germany

Luisa Vervier Human-Computer Interaction Center, RWTH Aachen University, Aachen, Germany

Marian Wiesch Laboratory for Machine Tools and Production Engineering (WZL), RWTH Aachen University, Aachen, Germany

Martina Ziefle Chair of Communication Science, RWTH Aachen University, Aachen, Germany

How Digital Shadows, New Forms of Human-Machine Collaboration, and Data-Driven Business Models Are Driving the Future of Industry 4.0: A Delphi Study

Frank T. Piller and Verena Nitsch

Abstract Transferring the idea of the Internet to the manufacturing landscape—the Internet of Production (IoP)—fundamentally changes our understanding of how products are developed, produced, and utilized. A key concept of the IoP is digital shadows that connect data, products, and equipment and are shared in cross-organizational data spaces. These developments are also core ideas driving the evolution of the current Industry 4.0 paradigm into its next generation ("Industry 4.U") and have far-reaching implications that go beyond mere technical issues. From a company-internal perspective, managers and workers need to deal with new forms of collaboration and cooperation between humans, robots, smart machines, and algorithms. From a company-external (network) perspective, data-based value creation and capture in platform-based ecosystems change the logic of many manufacturing business models. These changes have been reinforced by the COVID-19 pandemic, which acted as a catalyst for many transformation processes. Given the high uncertainty in the likelihood of occurrence and of the technical, economic, and societal impacts of these concepts, we conducted a technology foresight study in the form of a real-time Delphi analysis to derive reliable future scenarios featuring the next generation of manufacturing systems. This chapter introduces the conceptual and technical background of this study, defines important terms and frameworks, and provides an overview of the Delphi projections that are presented and analyzed in greater detail in the remaining chapters of this book.

F. T. Piller (✉)
Institute for Technology and Innovation Management, RWTH Aachen University, Aachen, Germany
e-mail: piller@time.rwth-aachen.de

V. Nitsch
Institute of Industrial Engineering and Ergonomics, RWTH Aachen University, Aachen, Germany

Fraunhofer Institute for Communication, Information Processing and Ergonomics FKIE, Wachtberg, Germany
e-mail: v.nitsch@iaw.rwth-aachen.de

[Abstract generated by machine intelligence with GPT-3. No human intelligence applied.]

1 Industrial Value Creation After the Pandemic

The COVID-19 pandemic has challenged politics, society, and the economy to an unprecedented extent. Although we have not overcome its consequences, it is apparent that the pandemic has acted as a catalyst, reinforcing existing trends, fundamentally changing our everyday economic life, and creating new structures (Piller et al., 2020). The crisis has given a major boost to digitalization in general and digital business models in particular. In response to the experience of the supply chain and labor shortages that followed the pandemic, companies in all industries will continue to automate their production even more and to transform their services into digital services more quickly. The business and societal models that prevail will be those that respond best to changing economic and social behavior and new societal demands. In this regard, the crisis has also demonstrated the importance of high-performing digital infrastructures and scalable communication networks for industry and healthcare systems and also in public administration and education (Agrawal et al., 2020).

At the same time, the crisis also revealed significant deficits and differences, particularly with regard to the digital maturity levels of various industries and sectors (e.g., with regard to real-time data processing capabilities, the maturity of digital processes, the speed of adaptation, and willingness to do so). It showed once again that those who succeed are those who are prepared. The pandemic demonstrated that the effects of such a global event are so drastic that they compel companies and industries not only to manage the crisis in the short term but also to develop strategic options for the future—to act proactively and not just react to and cope with new realities (Teece et al., 2020).

Against this background, an expert group from RWTH Aachen University, working together in a national Cluster of Excellence, *The Internet of Production* (funded by the German Research Council, DFG, as part of the German Excellence Initiative), set out to develop and validate a set of propositions on the future of industrial production with a projection horizon of 2030. Our objective was to create a picture of the future consisting of the elements of a next-generation production system that may exist in 2030. Such a picture of a desirable future can allow a backcasting process (Drehborg, 1996), i.e., working backward from the future scenario to identify policies and programs that would connect that specified future to the present and asking "if we want to attain a certain goal in 2030, what actions must be taken to get there?" (Holmberg & Robèrt, 2000). The results of our inquiry are presented in this book.

Our research was guided by the question of how the developments of the Fourth Industrial Revolution, or Industry 4.0, will evolve between now and 2030: what is the future of industrial value creation? First, we used a truly interdisciplinary panel of researchers, many of them co-authors of chapters of this book, to develop a set of projections for Next Generation Manufacturing, drawing on our own research in the cluster and also incorporating the inputs of numerous external professionals in the form of in-depth expert interviews. Using a novel real-time Delphi approach (Gordon & Pease, 2006; Gnatzy et al., 2011), we then validated these projections with the help of a large, international set of experts from multiple fields, e.g., engineering, information systems, social sciences, and management, who we asked about their evaluation of, and also their qualitative feedback and commentary on, the projections (with a projection period of up to 2030). After analysis, we consolidated the validated projections into different scenarios for the future of digital manufacturing.

Overall, we hope that our research will contribute to a more proactive design of the future of manufacturing. Our results can support both firms' strategic planning and future research. In particular, our study makes three major contributions:

- First, we provide a set of 24 validated projections regarding the future of digital manufacturing (with a projection period of up to 2030), based on 1930 quantitative estimations and 629 qualitative arguments from 35 industrial and academic experts from Europe, North America, and Asia. In so doing, we deliver a basis on which to substantiate academic discussions and which can support firm decision-making on future technological developments and economic implications that go beyond current speculations and siloed research.
- Second, we describe each projection in detail, offering current case study examples and related research, as well as implications for policy makers, firms, and individuals (managers, employees). These detailed projections can be used as a starting point for further research, but also for concrete strategic implementation in companies.
- Third, our empirical results allowed us to build scenarios for the most probable future along the dimensions of governance, organization, capabilities, and interfaces from both a company-internal and an external (network) perspective. In addition, we discuss emerging tensions between the internal and external scenarios. These scenarios will support managers when drafting new strategies and challenging those already in place.

While we are aware that a black swan event like another global pandemic (Taleb, 2005) may make some of our results obsolete, we believe that the research presented in this volume will prevail. We truly hope that nature (or humankind) will not tempt our fate in the decades to come.

2 Next Generation Manufacturing and the Internet of Production (IoP)

In this section, we (very) briefly review the evolution of manufacturing paradigms from the early stages of mechanization, through industrialization and digitalization, and into the current state of Industry 4.0. Building on this status quo, we describe our understanding and vision of the next generation of manufacturing (with a projection period spanning the decade from 2020 to 2030), which was our frame when developing the projections for the Delphi study. We use the term **Next Generation Manufacturing** to describe this envisioned state. This vision was informed by the research conducted by the author team of this book in the Cluster of Excellence "Internet of Production" at RWTH Aachen University, which we also outline in this section, explaining its central technical and structural elements. These were the starting points when developing the projections of our Delphi study. Our focus is the *usage stage*, i.e., operating a future digital and networked manufacturing system, from both an internal perspective (user acceptance, work organization, human-machine interaction) and an external perspective (industrial business ecosystems and data-driven manufacturing platforms). Finally, we introduce the five factors that structured our Delphi analysis (governance, organization, capabilities, interfaces, and resilience).

2.1   From Industry 4.0 to Industry 4.U

The evolving Fourth Industrial Revolution is currently reshaping manufacturing industries through the broad deployment of new digital manufacturing technologies, but also new digital business models driven by these technologies. The term "Industry 4.0" has been established as a metaphor for this fundamental change in the way in which we organize production and value creation. The term was popularized in Germany ("Industrie 4.0") after being coined at the Hannover Messe Industrie (HMI) in 2011.

The advent of mechanical machinery and the steam engine triggered the First Industrial Revolution, facilitating the change from hand-crafting methods to industrialized production. A key development here was the first mechanical weaving loom (1784). The Second Industrial Revolution started with the rise of mass production systems, building on the first meatpacking assembly line (1870), further driven by Taylor's "Principles of Scientific Management," and with Ford's conveyor belt as its enabling technology. In the late 1960s, programmable controllers paved the way to computer-aided manufacturing and planning, starting the Third Industrial Revolution but also marking the beginning of the continuous digital transformation of manufacturing. A core technology here was the first programmable logic controller (PLC) in machinery (the Modicon 084, 1969). The spread of graphical user interfaces and the networking of individual machines to form larger production islands further promoted the computerization of production. Since the 1980s, when the term computer-integrated manufacturing (CIM) became a buzzword for new production concepts, AI-driven computer networks have been envisioned as revolutionizing the future of production. Radiofrequency identification (RFID) technologies (1999) further enabled flexible automation scenarios in production.

The introduction and spread of new technologies have always been accompanied by a change in work activities, conditions, and forms of organization. With the development and introduction of CNC machine tools, the first industrial robots, and production lines, human work was often replaced by automation or reorganized according to Tayloristic principles into highly specialized, repetitive, often physically demanding activities (Schlick et al., 2018). With the introduction of the first database systems and user-friendly dialog processing systems, it was later possible to combine sub-functions at workstations, thus creating more varied work profiles which also often demanded more qualifications. With the rapid advance of personal computers, knowledge work gained in importance and attention, and today humans largely create value by employing knowledge rather than muscle power and coordination. Knowledge work is specifically characterized by a high degree of autonomy and a certain degree of (result- and process-related) uncertainty; it is complex, is communication-intensive, and includes routine as well as creative activities, which distinguishes it from traditional skilled work (Wilkesmann, 2005). This type of work places fundamentally different demands on human's physical and cognitive abilities, and thus, the requirements of modern production technologies that aim to enable knowledge workers are changing accordingly. New technologies also enable more flexible ways of working (e.g., Ahlers, 2016). For instance, working from home and virtual teamwork have only become possible through the location-independent availability of information and applications, which is made possible by digitalization.

The technology driving the Fourth Industrial Revolution (Industry 4.0) is a unifying network infrastructure connecting human actors, machines, and products. This Internet of Things (IoT) has transferred the original Internet (of communications) into the physical world (Gubbi et al., 2013). Hence, in the US literature, the terms Industrial Internet and Industrial Internet of Things (IIoT) are used with similar connotations. They refer to a subset of the general digital transformation of existing businesses and processes in which digital structures replace previously analog or even manual operations (Sisinni et al., 2018; Porter & Hemppelmann, 2015). The technical enablers of this development are cyber-physical systems, which integrate smart devices with sensing, communications, network, and autonomous acting capabilities (Dalenogare et al., 2018; Porter & Heppelmann, 2015; Reischauer, 2018). These systems' technology stack consists of a classical device layer, i.e., the physical device and the added logical capability of embedded sensors and actuators, a network layer for the transmission and transport of information, a content layer that contains the data and metadata, and a service layer for the application functionality (Fleisch et al., 2014). The real-time data stream can be analyzed for decision-making purposes and to control devices flexibly throughout the entire value generation process (Hartmann & Van der Auweraer, 2021).

While this development generated vast amounts of data, most of it is currently stuck in silos—it is neither easily accessible, nor interpretable, nor connected to knowledge gain (Schuh et al. 2020). In addition, many concepts from dealing with consumer data are not transferable to manufacturing contexts, which are characterized by many more parameters but often much less available data. Hence, data on the usage of products is not yet used across manufacturing firms to optimize operations, investment decisions, innovation processes, or the generation of new products.

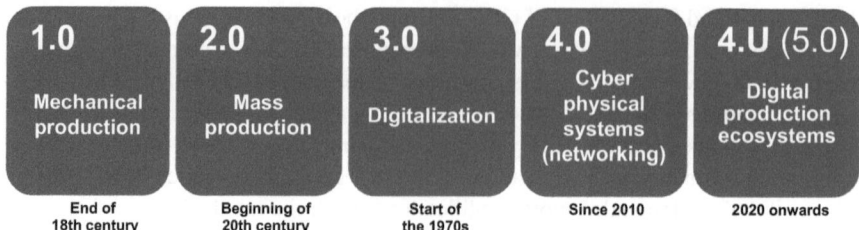

Fig. 1 The evolution of industry: from the first mechanical weaving loom (1784) via the first meatpacking assembly line (1870) and the first programmable logic controller (PLC) (the Modicon 084, 1969) + RFID (1999) toward Industry 4.0 (HMI 2011) and Industry 4.U (Internet of Production, 2020)

Shared data is also a core driver of sustainability in manufacturing (Bai et al., 2020; Ghobakhloo, 2020). Data-enriched views on processes and increasing information capability (in real time, complete, distributed) are the underlying principles of improving the efficiency of processes and avoiding waste. Our vision of *Next Generation Manufacturing* (derived from our research in the Internet of Production research cluster) is based on this background of shared data in inter-organizational production ecosystems. One of its main enablers is cross-company data spaces, which link classic data silos across organizational boundaries (Otto & Jarke, 2019; Cappiello et al., 2020). Supported by artificial intelligence and machine learning, new insights can be created. Data spaces can be seen as one form of business ecosystem, changing a firm's business model as a result of a digitalized and networked manufacturing system (Otto et al., 2019). This vision also resembles the description of the next generation of digital manufacturing systems outlined by ElMaraghy et al. (2021).

Based on an idea by our colleague Thomas Gries (2020), we suggest calling this further evolution of the Fourth Industrial Revolution "**Industry 4.U** (for you)," which recognizes that Next Generation Manufacturing systems and the business ecosystems they are built upon are focused on creating ultimate value for customers, but also for our society and the planet. This purpose complements the current perspective of Industry 4.0, which has been predominantly focused on increasing operational efficiency (OE, OEE). Industry 4.U builds on open, cross-organizational business ecosystems, which generate value for customers and society and enable novel business models with the ambition of ecological and social sustainability in manufacturing (see Fig. 1).

The central purpose of Industry 4.U is to increase value for customers and users. New supply chain structures with flexible processes and high equipment efficiency not only deliver cost savings but also enable a range of strategic benefits such as better handling of complex products, short time-to-market, and on-demand manufacturing. Connecting and sharing data in open business ecosystems yields new value propositions for highly customized or differentiated products, well-synchronized product-service combinations, and value-added services (Burmeister et al., 2016). While differentiation and cost leadership have conventionally been considered contradictory strategies, Industry 4.U promises to enable both simultaneously. Shared data also increase a firm's ability to predict demand by analyzing different streams of market information. This allows more precisely timed procurement and producing only what is needed, when it is needed—providing customers with what they want, when they want it, but also reducing waste from overproduction at all levels of the value chain.

The latter result is part of the second purpose of Industry 4.U: utilizing manufacturing ecosystems and cross-organizational data spaces to realize vast new opportunities for more sustainable production. Sustainability may actually act as a driver of structural change: in parallel with our current digital transformation, we need a sustainability transformation of today's value creation and production models into future-proof, resilient approaches (Piller et al., 2022). But reducing resource consumption and making existing production models more efficient are just the beginning. CO2-neutral production and the shift from energy management to resource efficiency are accepted goals in many companies today. However, these measures—while undisputedly important—increase path dependencies and focus on existing business models, often supporting only a sustainability narrative ("green- or social-washing"). Instead, a true triple paradigm shift is required: digitization, sustainability, and customer-centricity either must move to the center of the value proposition or must be generated by a digital value creation structure itself.

Next Generation Manufacturing is also accompanied by a paradigm shift from technology-centered toward human-centered digitalization and work design (cf. Mütze-Niewöhner et al., 2022, Hirsch-Kreinsen and Ittermann 2021). Most would agree that CIM failed because it paid no attention to workers (Thorade, 2020). Industry 4.U aims to learn from the mistakes of CIM by consistently considering the role of humans in the socio-technical work system when developing, introducing, and implementing Next Generation Manufacturing technologies in workplaces. Thus, Industry 4.U no longer aims to use automation technologies to replace humans as comprehensively as possible, but to support them at work in an individually customized manner by taking their individual capabilities, habits, and preferences into account. By providing adaptive, individually optimized support, technology is considered a key enabler of productive, healthy, and safe work. In the future of production, technology will be used to support socially innovative, economic, and humane work design, thus maximizing social sustainability at work.

2.2 The Cluster of Excellence "Internet of Production (IoP)"

The context of this work is the interdisciplinary research cluster *Internet of Production (IoP)* at RWTH Aachen University (iop.rwth-aachen.de). RWTH Aachen is characterized by a unique scope and an outstanding reputation in production research. An earlier Cluster of Excellence, *Integrative Production Technology for High-Wage Countries* (2006–2017), was awarded to RWTH Aachen as part of the Excellence Strategy inaugurated by the German federal government to strengthen its university landscape in 2005. This first cluster focused on the development of innovative solutions to ensure the future viability and competitiveness of the local manufacturing industry. Achievements included, for example, the development of new intelligent production systems, solutions for the efficient production of customer-specific components, integrated product life cycle management (PLM), and predictive human-robot collaboration (HRC) concepts, as well as increased interconnectedness and collaboration, laying important foundations for the development of the vision of Industry 4.0.

In 2017, research started in the following Internet of Production cluster. Since then, about 100 researchers at all career levels and from various disciplinary backgrounds, such as mechanical and plastics engineering, material science, industrial engineering and ergonomics, humanities, management, and computer science, have been working on a vision for a new level of cross-domain collaboration along the entire product life cycles, from engineering to operations and usage (Brecher et al., 2016). The IoP pursues a vision called the World Wide Lab (WWL), in which processes, factories, and even organizations can learn from each other by sharing experiences and knowledge (Brauner et al. 2022). Corresponding to the relationship between the Internet and the World Wide Web (WWW), the WWL aims to be a network of multi-site labs in which models and data from experiments, manufacturing, and usage are made accessible even across company borders to allow companies to gain additional knowledge. This change will increase productivity in a similar way as the WWW increased the efficiency of e-commerce transactions, customer interactions, and entertainment.

A main driver of the WWL are task- and context-dependent, purpose-driven, aggregated, multi-perspective, and persistent datasets, which we call *digital shadows* (Bauernhansl et al., 2018; Riesener et al., 2019). Digital shadows are multi-modal views with task-specific granularity which can simultaneously provide high performance, low latency, security, and privacy (Brauner et al. 2022). They enable radically new kinds of production and engineering applications. The cross-domain exchange of digital shadows in the form of *data spaces* can make data more valuable by opening up the current data silos of different companies to increase the speed of research and innovation in the presence of global challenges such as the three Ds (demographic change, digital transformation, and de-carbonization). The following section will explain the concept of connected digital shadows in more detail.

The Internet of Production strives to integrate the major domains of a manufacturer along the life cycle of a product (development, production, and usage). This integration, enabled by the infrastructure of the Internet of Production, results in a new level and understanding of cross-domain collaboration due to the real-time availability of semantically adequate and contextual data from these domains. The participating research groups thus approach the Internet of Production from different application perspectives, taking up the challenge of bringing together methods from material science, production engineering, and production management, along with human factors and business models. Cross-domain research and cooperation with industry are coordinated by a set of use cases at different scales, ranging from rather technical process innovations to complex, life cycle-wide interplays of subtasks to inter-organizational network design (Liebenberg & Jarke, 2022). For this purpose, the RWTH Aachen Campus (rwth-campus.com), with its various research institutes and industrial partners, offers unique infrastructural opportunities for the integrative development and validation of the Internet of Production.

2.3 Digital Twins and Digital Shadows

As introduced before, the sharing of knowledge, models, and data across all relevant domains within and between manufacturing firms is part of the value proposition of the next generation of Industry 4.0 (Björkdahl, 2020). A key enabler of our vision of Next Generation Manufacturing is *digital shadows*: purpose-driven, aggregated, multi-perspective, and persistent datasets from production, development, or usage (Liebenberg & Jarke, 2020). Digital shadows are a specific subset of the broader idea of *digital twins*, one of the core elements of Industry 4.0. Digital twins are a "description of a component, product or system by [a] set of well aligned executable models [. . . linking] engineering data, operation data and behavior descriptions [. . .] along the whole life cycle" (Boschert, Heinrich, and Rosen, 2018). A more practice-focused description of digital twins has been provided by the Digital Twin Consortium, an industry group (digitaltwinconsortium.org). According to their definition, a digital twin is a virtual representation of real-world entities and processes, synchronized at a specified frequency and fidelity. They facilitate holistic understanding, optimal decision-making, and effective action of manufacturing systems. Digital twins use real-time and historical data to represent the past and the present and to simulate predicted futures. They are motivated by outcomes, tailored to use cases, powered by integration, built on data, guided by domain knowledge, and implemented in IT/OT systems.

When integrated, a digital twin allows objects (end products, components, machinery, infrastructures, etc.) to be tracked and controlled throughout their entire life cycle, facilitating a continuous adaptability to changing customer requirements. For example, digital twins enable manufacturing firms to shift from acquiring reliable industrial equipment to paying for access to outcome-based services and performance through the data generated by the equipment (Iansiti and Lakhani, 2020; Porter and Heppelmann, 2015). In such a scenario, digital twins enable the provider of the machinery not just to measure the volumes of products or components manufactured on that machine but also to enable production without unplanned downtime and with continuous improvement.

Hartmann and Van der Auweraer (2021, p. 7) recently suggested the idea of an *executable digital twin*, "a specific encapsulated realization of a digital twin with its execution engines . . . [which enables] the reuse of simulation models outside of R & D." To realize this objective, a digital twin will be purposefully designed for a specific use case or industrial application, using existing data and models. The vision is that "an executable digital twin can be instantiated on an edge, on premise, or in the cloud and used autonomously by a non-expert or a machine through a limited set of specific APIs" (Hartmann and Van der Auweraer, 2021, p. 7).

In practice, the challenge is the integration of the different scales (temporal, spatial, etc.) of the numerous underlying processes, which yield huge amounts of data, ill-fitted models, and high latencies if data needs to be aggregated and analyzed. Even in a mid-term perspective, we do not consider a complete digital twin to be feasible due to the massive amounts of data that a virtual replica of a product, machine, or production plant updated with a high frequency would require. Also, most digital twins used in practice are not complete digital counterparts of physical objects; rather, they are collections of different datasets and models, each representing a particular aspect of the real object. Hence, a core element of our vision for the Internet of Production is the *digital shadow*, which we consider as a collection of task- and context-dependent, purpose-driven, aggregated, and persistent datasets that encompass a complex reality from multiple perspectives in a more compact fashion and with better performance than a fully integrated digital twin (Brauner et al., 2022; Schuh et al., 2020). Figure 2 provides an illustration of the different levels of digital shadows in a production system. A digital shadow can be compared to a view in database systems: an aggregated subset of the data on the real object computed by a complex function that might include complex algorithms for data reduction and analysis (Becker et al., 2021).

In our understanding of the Internet of Production, digital shadows are the "units of data" shared among organizations. They connect data, products, and industrial assets within and across organizations and are the foundation for data-driven planning and decision-making within an organization (factory) and between organizations (supply chains, value chains). Hence, digital shadows have company-internal and company-external implications in Next Generation Manufacturing:

- Internally, digital shadows enable different forms of production organization by enabling, for instance, changing human-machine interactions in the context of collaborative robotics, remote work in production contexts, or artificial intelligence-based decision support systems.
- Externally, integrating user and usage data into a digital shadow creates (open) data ecosystems which enable new forms of collaboration and innovation.

We will discuss these internal and external implications in more detail in the following two sections.

2.4 Internal Perspective: New Forms of Human-Machine Collaboration Enabled by Digital Shadows

Internally, (connected) digital shadows influence how manufacturing is organized and provide new opportunities for human-machine collaboration. In a Next Generation Manufacturing system, interactions are not only carried out between

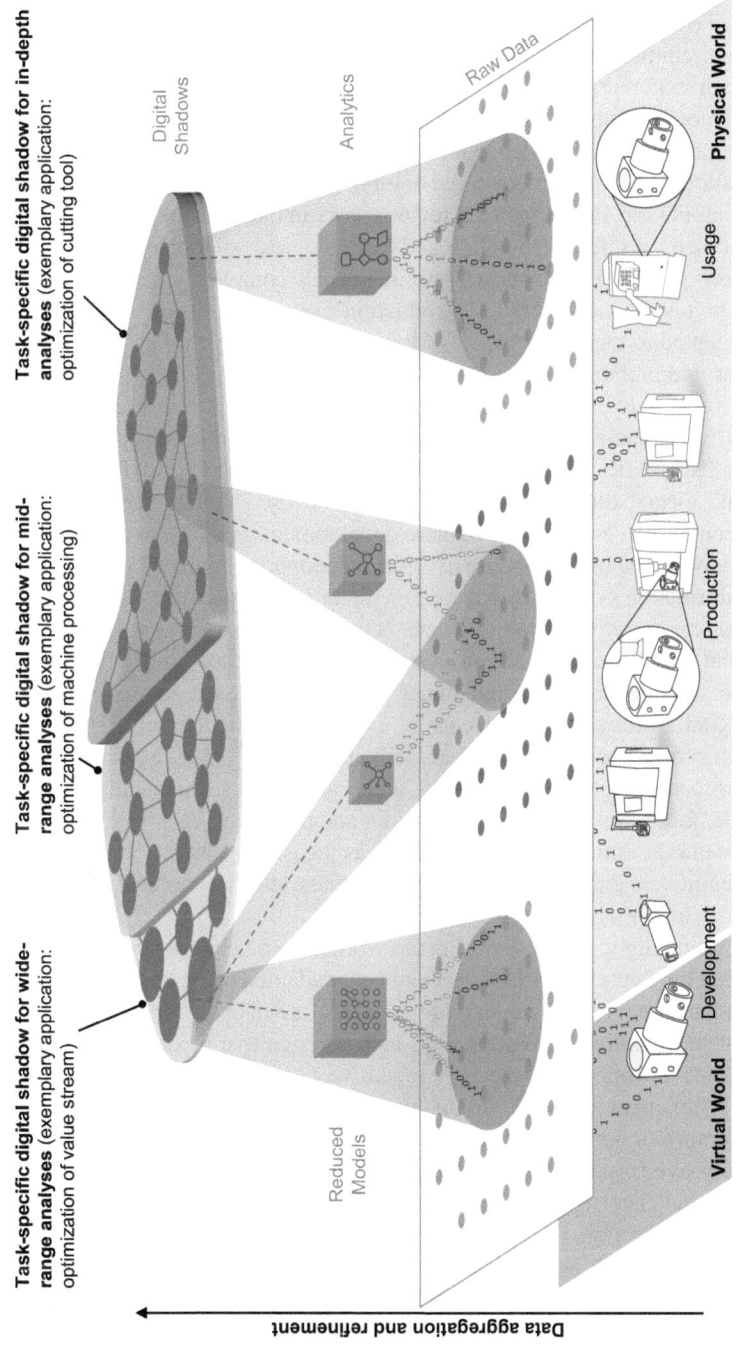

Fig. 2 Network of task-specific digital shadows for wide-range, mid-range, and in-depth analysis (Source: iop.rwth-aachen.de)

autonomous machines. Through intelligent connectivity and the target group-specific provision of information, machines learn from humans and vice versa. Irrespective of their age, gender, and educational, cultural, and social background, humans will continue to be an essential part within future manufacturing systems, even though concrete tasks, qualification requirements, and work structures may change. Hence, an important task is to design systems for user-centered communication, interaction, and knowledge transfer between persons and digitalized production technology. In human-robot collaboration, the advantages of both manual work and high automation are combined. Previous work in industrial application scenarios mainly focused on safe collaboration and technological solutions to avoiding traditional safety guards. But to master the increasing complexity of the information available in future socio-technical production systems, decision support systems must be adapted to the requirements of human operators, including their individual capabilities and preferences.

Novel forms of hybrid teamwork in terms of joint actions, e.g., predicting human intentions in human-robot collaboration, and an appropriate mapping of human skills and capabilities to the technical systems shape the basis for an efficient interaction. Virtual training environments can be used to familiarize the user to complex contexts. A core enabler of these developments could be to complement the rather techno-centric understanding of digital shadows (stemming from an engineering and information systems perspective) with an anthropocentric perspective in the form of *human digital shadows* (Mertens et al., 2021).

A human digital shadow comprises all data that can be assigned to human actors within the socio-technical system under consideration as a source or sink. The human digital shadow thus represents an extension of the idea of digital twins or shadows of products or machinery but facilitates the analysis of existing and future interactions *between* people, technology, and organizations. Human digital shadows can include, for example, patterns of behavior and movement of a worker, individual working methods, anthropometric data such as measurements or forces, physiological and cognitive parameter progressions and states, abilities, skills and experiences, and other socio-demographic information.

In the future, digital shadows of human workers can help ensure that the interaction between humans and machines is even better adapted to the user and the context, e.g., in human-robot collaboration. Humans therefore take on context-specific tasks that match their abilities and strengths, such as cognitive flexibility and situation awareness, while robots execute very repetitive tasks that require high speed and precision (Giuliani et al., 2010). In this use case, the application of a human digital shadow comprises both human-robot cooperation, i.e., sharing a workspace, with occasionally overlapping work, and human-robot collaboration, i.e., working hand in hand with mandatory physical contact to fulfill the assigned task. In a basic case of human-robot interaction, the human performs a task, while the robot assists, e.g., by holding and handing tools or lifting, positioning, and machining work pieces that the human is actively working on (Mertens et al., 2021). Hence, the robot acts similar to a colleague supporting the human worker. This interaction could be enhanced using the data from the digital shadow of a human, e.g., data pertaining to individual working methods and procedures, handedness, or his or her experience and expertise. It could become possible to simulate mechanisms to aid familiarization that would otherwise typically occur on a social level and as part of interpersonal interactions. Thus, a human digital shadow can contribute to both increased efficiency and effectiveness in operations and higher levels of acceptance of robot support.

By capturing the tacit knowledge and experience of workers, human digital shadows could also facilitate company-wide knowledge management, enabling the replication of practical know-how in a novel way that could not be realized before. If solutions are found to be particularly beneficial for specific tasks, these could be transferred to other workplaces, and technical assistance systems could apply them when applicable. In this way, other human workers could implicitly assimilate and benefit from the insights of others during their own work without the acceptance problems that often arise during explicit training (Mertens et al., 2021).

Human digital shadows could also facilitate better planning and decision-making in Next Generation Manufacturing systems. Despite all the advancements in AI and autonomous systems, it is our strong belief that humans will continue to perform tasks of information acquisition, information analysis, and decision-making in production (Parasuraman, Sheridan, & Wickens, 2000). Due to the rapidly advancing digitalization of production, humans will increasingly be supported by decision support systems and other technical assistance systems to accomplish those tasks. Thus, an assistance system could indicate which component to pick next, visualize the current allocation of workers, or simulate the outcome of reorganizing an assembly line or supply chain. In these scenarios, task performance depends on the assistance system's ability to support the user, which, in turn, depends on how well the system adapts to the specific needs of the user and the demands of the task (Mertens et al., 2021). Applying human digital shadows to the system would provide opportunities to significantly improve such collaborative systems. By modeling a user's past interactions with the system (based on his or her digital shadow), individual assistance tailored to his or her mental model could be provided. In a further step, the human digital shadow could also incorporate more advanced data, including an assessment of the user's mental state during task execution, e.g., cognitive load, mental fatigue, and attention allocation. Valid indicators for relevant human states and well-being which use psychophysiological measures such as pupil dilation, heart rate variability, or skin conductivity already exist and can be implemented contactlessly or, e.g., with the help of a smartwatch (Charles & Nixon, 2019).

These outlined applications are just a few of the scenarios in which connected digital shadows of humans and machines may influence the next generation of manufacturing. One of the central advantages of human digital shadows is that problems cannot only be prevented, but resources can also be used more efficiently in the short term. Technical changes could support this goal: for instance, equipment, tools, and user interfaces could be adapted or even customized to each individual worker. However, we strongly believe that the introduction and use of digital shadows of humans should always be guided by a proper understanding of the potential ethical risks and accompanied by strong efforts to mitigate against these risks. Without addressing these concerns, neither compliance with established ethical standards nor employee acceptance will be achieved.

2.5   External Perspective: Data-Driven Ecosystems Creating and Capturing Value from Digital Shadows via IoT Platforms

Externally, integrating engineering, production, and usage data in the form of digital shadows is the underlying foundation of new, data-based industrial ecosystems. Our vision for the Internet of Production is, as its name indicates, not restricted to a focal company or value creation within a closed network of established partners. It resembles an open network of sensors, assets, products, and actors that continuously generate data (represented in digital shadows). These data are utilized to enhance operational efficiency, but also to provide new opportunities for strategic differentiation. A core element is thus the (re-)usage of data, insights, and applications by other parties than those generating the data in the first place.

In the management literature, this idea corresponds to the shift from conventional value chains to *platform-based business ecosystems* which mediate data and connected assets with third-party complements (Kopalle, Kumar, and Subramaniam, 2020). Since the early 2000s, the industrial organization literature has begun to develop theory on platforms, also referred to as "two-sided markets," "multi-sided markets," or "multi-sided platforms" (Rochet & Tirole, 2003). Economists view platforms as special kinds of markets that play the role of facilitators of exchange between different types of users that could not otherwise transact with each other. The rise within industrial ecosystems of platforms where these data are being exchanged and enhanced by dedicated "apps" (complementing offerings and services), often provided by specialized third-party entities, is one of the largest economic developments of the last decade (Adner, 2017; Gawer, 2014).

Platforms connect multiple sides to enable transactions or foster innovation. They can be considered as the technological architecture on which firms develop platform-based business models (van Dyck et al., 2021). For that, they need to coordinate the network of users (customers, consumers) and providers (complementors). Together, they build an ecosystem consisting of a central platform with multiple peripheral firms connected to it (Gawer, 2014). Following the dominant view in the literature, platform-based business models are in most instances orchestrated by a central (keystone) player (Adner, 2017).

In the context of industrial manufacturing, the term *IIoT (Industrial Internet of Things) platform* is frequently used to denote such a platform. As Fig. 3 illustrates, they have to be differentiated from platforms in the form of marketplaces focusing on the transaction of goods (transaction platforms, such as Amazon or Alibaba) or data (*data spaces*, such as IDS or GAIA-X). IoT platforms are the technical

Types of BtoB Platforms		
IIoT Platforms	**Data (Transaction) Platforms**	**Transaction Platforms**
Connecting processes, assets, and machines for automated interaction of these "things" and enabling analytics, predictions, and prescriptions about them through dedicated (3rd party) applications	"Data spaces" for exchange and monetarization of relevant business data from all domains (production, product, weather, etc. data) across different organizations	Platforms with transaction focus for exchange and trade of materials, components, and products between actors in a unified digital environment (marketplaces, logistic, recycling networks)

Fig. 3 Types of business-to-business platforms

implementation of the Internet of Things (IoT), which, as defined before, aims to connect physical and virtual objects and enable them to work together. The spectrum of such so-called smart devices ranges from household appliances to transportation and logistics systems to industrial plants. Networking these devices in practice means connecting them to an IoT platform. The platform fulfills the function of an operating system that enables application programs to read data from the various devices and send control signals to them using standardized Internet technologies (Guth et al., 2016; Hoffmann, et al., 2018).

IoT platforms are thus an essential prerequisite for the connected devices to provide actors with greater benefits through innovative applications than the devices are capable of themselves. Generally, learning and analytics can take place faster and more efficiently if manufacturers not only utilize their own data but can also access data from similar contexts in other industries. Hence, in the 2010s, more and more de facto standards (e.g., the RAMI 4.0 initiative) emerged to facilitate greater connectivity and networking via different IIoT platforms. These standards enable a wide range of complementors to develop applications for one IIoT platform. Users of platforms with open interfaces (i.e., APIs based on non-proprietary standards) have the advantage of being able to select each application independent of the manufacturer according to individual requirements. The more open an IIoT platform is, the lower the transaction costs and investment risks associated with its introduction. Because third-party providers can also access an open API, open IIoT platforms favor a broader spectrum of providers and the development of innovative new applications (Guth et al., 2016).

In the case of Next Generation Manufacturing, IIoT platform participants include a number of entities:

- The *orchestrator of the platform* is today either an IT infrastructure provider like SAP, Microsoft, or Amazon Web Services (AWS); a provider of automation or manufacturing equipment like Siemens, GE, or Trumpf; or, in a recent development, an industrial producer, like Volkswagen Group with its Industrial Cloud platform.

- *Users of the platform* are operators of production assets (users in the form of "factories"), which provide data and receive analytics, predictions, or prescriptions. Some operators only utilize insights from aggregated data to optimize their own production (in exchange for a fee), but do not share any of their own data. Other users also become . . .
- *Providers of applications* ("app providers"), who develop applications for analyzing data and providing prescriptions and predictions. App developers can be specialized firms (e.g., data analytics startups, research institutes, but also established analytics providers) or lead users (plant operators) who share applications originally developed for their internal usage with other members of the platform ecosystem (basically complementing their existing business model with another one).
- In addition, the goods being produced can also become part of the platform in the form of *connected ("smart") products*, providing a feedback loop of usage data and also becoming the center of another platform based around digital services complementing these products. The latter also refer to end-users (consumers) as a final participant of the platform.

An essential prerequisite for making multi-sided IIoT platforms work is the existence of network effects created by these participants and arising between the different sides of the market (Gawer, 2014). As the value of the platform stems principally from the access of one side of the platform to the other side of the platform (e.g., users providing data, users providing analytics apps, and users paying for the insights generated from these data analytics), the question of platform adoption becomes one of how to bring multiple sides on board. To create value, platform-based manufacturing ecosystems hence depend on complementary inputs made by loosely interconnected yet independent stakeholders with varying levels of (technological) distance from each other and the end consumer.

This will not just change the business model of an individual company but will create entirely new industry structures. As a result, manufacturing firms may change roles from producer-sellers to platform orchestrators (van Dyck et al., 2021). This not only challenges traditional business-to-business relationships in highly vertically integrated, rigid supply chains characterized by asset- and process-specific investments (Sjödin et al., 2016) but also requires a different form of value creation in which multiple sides of the market are connected and firms compete through the data generated by products (Kopalle et al., 2020). While some firms might profit from such an ecosystem approach, competitive advantage may shift from the machine (hardware) to the data (digital shadow) layer. This could have detrimental effects for firms without continued data access (e.g., Alexy et al., 2018; Dahlander, Gann, and Wallin, 2021). Therefore, incentives, governance, and new ways of user integration are necessary elements to make this vision a reality (Kortmann & Piller, 2016).

Hence, a core element of Next Generation Manufacturing systems is mechanisms of ecosystem (platform) governance (Adner & Kapoor, 2010). Dedicated mechanisms governing data sharing and access are required to avoid the misuse of data. At the same time, appropriate incentives must be set in order to align the different interests and priorities of the partners involved (in order to encourage the sharing of data (digital shadows) in data spaces and other forms of exchange). In addition, managing property rights (access, transfer, enforcement) regarding data, applications, and connected assets will become a core capability of manufacturing companies, which will also need to define governance modes and design factors in order to generate adequate business models that allow value appropriation to be maximized by all involved actors. In such a situation, a core decision for a platform operator is how open to make the platform and when to absorb inputs (developments, apps, data) from the connected parties (Parker & Van Alstyne, 2018).

Currently, we see these dynamics unfolding in parallel in many industries. We are far from a consolidation of industrial platforms, as has happened in the consumer markets of social media, mobile phone operating systems, ride sharing, or entertainment. But these examples also indicate the profound effect the rise and success of one platform orchestrator can have. One of the central intentions of our Delphi study was to provide more clarity by developing validated forecasts for future data and application platforms in industrial ecosystems over the coming decade.

3 Strategic Design Factors of Next Generation Manufacturing: A Framework for Analyzing the Delphi Projections

The previous sections introduced the fundamental elements of a Next Generation Manufacturing system, building on the vision of the Internet of Production. Utilizing connected data spaces with digital shadows across organizational borders can yield new dimensions of industrial value creation when the right interfaces are designed, the required capabilities are developed, a specific organizational structure is established, and an adequate governance mode, accepted by all members of the ecosystem, is adopted. In this section, we present the analytic framework used to develop our Delphi projections for Next Generation Manufacturing and to analyze the responses from our Delphi panel. Following Gawer (2014) and Parker and Van Alstyne (2018), we distinguish four strategic design factors which influence the utilization and exploitation of digital shadows and data spaces internally and externally: governance, organization, capabilities, and interfaces. The development of the projections for our Delphi study can be structured along these four dimensions:

- *Governance* refers to the impact of digital shadows and cross-organizational data spaces on business models and governance structures in manufacturing organizations. A particular focus is on the level of value capture, i.e., how the different actors of an IIoT platform can profit from their participation and contributions to the platform. This dimension also considers the rules and regulations for data exchange across organizational boundaries.
- *Organization* studies the effects of connected digital shadows and AI on work organization and human-machine interactions. Who is making the decision:

humans, machines, or both in a hybrid, collaborative mode, and how will this collaborative work be organized?

- *Capabilities* refers to the new skills and abilities organizations need to utilize the opportunities derived from digital shadows and collaborative work environments, but also to the capabilities that Next Generation Manufacturing systems offer for increasing efficiency and achieving real sustainability in industrial production.
- *Interfaces* address the different layers of interfaces between a human operator and a future manufacturing system, from both a cognitive and a spatial dimension. From an external perspective, the openness of machine-to-machine interfaces (APIs) is investigated as a key design factor of future (IIoT) platforms.

Influenced by our shared experiences during the COVID-19 pandemic, we added a fifth cluster of projections addressing the need for resilience in Next Generation Manufacturing systems.

- *Resilience:* next-generation production systems may utilize data-driven ecosystems based on collaborative applications and using leading AI technologies and Industry 4.0 standards to anticipate disruptions (anticipation) and to optimally adapt their production planning to mitigate against active disruptions at any time (response).

The following section elaborates on each design factor and the related projections tested in our Delphi study.

3.1   Governance

Technological change requires firms to reconfigure their business model in order to maintain a consistent alignment between external and internal factors (Snihur, Zott, and Amit, 2021). With digital shadows and data spaces mediating data sharing across organizations, interdependence among products and services from multiple firms is increased, as these jointly form a value proposition in a future manufacturing ecosystem. As a case in point, Siemens' MindSphere platform integrates data from connected products and plants and invites third-party developers to build customized applications using the data. Here, MindSphere both serves as a source for innovation and also orchestrates the interaction between app providers and users. Thereby, Siemens is able to tie its business model to the customer more closely, to include stakeholders previously not present in the industry, and to tap into what Hedenstierna et al. (2019) describe as economies of collaboration. As a consequence, activities across the value chain are redistributed, and customer interactions change. At the same time, designing and implementing a joint value proposition across organizations is contrary to traditional business-to-business relationships, which are characterized by rigid supply chains and hierarchical structures (Sjödin, et al., 2016). Business models for Next Generation Manufacturing thus also need to

Table 1 Projections for governance structures in Next Generation Manufacturing

P#	Short title	Projections for governance structures in Next Generation Manufacturing
1	Subscription Models	In 2030, subscription models for production machines will be the new industry standard, fulfilling an assured performance level based on real-time usage data in return for a periodic payment
2	Digital Services	In 2030, for production machinery and other hardware assets, e.g., tractors, equipment, etc., competition will shift from hardware capabilities and functionality to differentiation by (digital) services, supplementing the traditional transactional business logic with a data-driven business model
3	Data Sharing	In 2030, organizations that share usage and production data with suppliers, customers, and other partners will obtain a competitive advantage over organization that do not share this data
4	Central Platform	In 2030, one central platform provider will serve as the operating system for the Industrial Internet of Things, enabling them to make use of data by integrating machine manufacturers and complementary service providers and to capture the greatest share of the value created
5	Data Mediator	In 2030, platform orchestrators or dedicated third-party providers will mediate data sharing between all actors involved in a production network
6	Industrial GDPR	In 2030, industrial data protection regulations (like a special GDPR—General Data Protection Regulation for Business-to-Business) will govern the application of data-based digital services

focus on capturing sufficient proportions of the value created between all actors (Björkdahl, 2020).

However, envisioning new business models is particularly difficult in emerging ecosystems (Dattée et al., 2018), creating high uncertainties for established firms. Hence, the Delphi projections for this cluster (Table 1) will help firms to identify favorable business model elements and governance structures in light of developments in future digital manufacturing. A particular focus here is on the level of value capture, i.e., on mechanisms that allow the different actors of an IIoT platform to profit from their participation in and contributions to the platform. The first three projections of this cluster cover such advanced approaches to capturing value from manufacturing platforms (Projections P1, P2, P3). The next projection predicts a market concentration of IIoT platforms (toward a "winner-takes-all" situation), similar to the development of digital platforms in the consumer market (P4)—a development that probably would not be the best solution. Therefore, we also consider the rules and regulations for data exchange across organizational boundaries (P5, P6).

3.2 Organization

Given their central importance for organizing manufacturing, the design of human-machine interactions has been a core field of interest in research and practice alike.

Table 2 Projections for organization routines in Next Generation Manufacturing

P#	Short title	Projections for organization routines in Next Generation Manufacturing
7	Autonomous Robots	In 2030, collaborative robots that move autonomously on the shop floor and interact directly with humans will have replaced most conventional robots that only interact in protected cells
8	Hybrid Intelligence	In 2030, strategic production decisions will be executed with close interaction between humans and AI-based algorithms ("hybrid intelligence")
9	AI-Based Assistants	In 2030, operative production decisions will no longer lie with people, as they will be made by AI-based decision-making agents
10	New Leadership	In 2030, AI-based decision systems will have changed our current understanding of management completely, increasingly eliminating hierarchies and leadership based on human interactions
11	Human Digital Twins	In 2030, a full digital twin of each production worker and all of her/his operations will be available and will become a valuable tool for production planning and optimization by reflecting their workload, their stress, and also their need for training in real time
12	Employees' Rights	In 2030, adequate anonymization procedures for the protection of employees' personal rights will have been introduced for firms that collect data on personal performance and work patterns in the form of digital twins of their employees
13	Workforce Reduction	In 2030, AI-based software and robots will have reduced a company's workforce significantly

Similarly, for the next generation of manufacturing systems, interfaces between humans and machines will be the enablers of human workers' operational and strategic decisions and actions (Nelles et al., 2016; Shin, 2014). Task demands must correspond to operators' visual and cognitive ergonomic requirements in order to support efficient and responsible decision-making. Hence, we were interested in exploring a set of Delphi projections in this domain (see Table 2). In the usage stage (the focus of our Delphi study), the emphasis is on knowledge generation and improving decision-making by integrating human actors with technical systems (Brauner et al., 2022, Villani et al., 2017). In this context, usage refers to capturing the value of human and machine capabilities as an essential component of socio-technical production systems. With apps and AI augmenting the autonomy of humans in decision-making, one question often asked is who is making the decisions in the factory (network) of the future: humans or machines?

Interaction between the entities of a socio-technical production system leads toward hybrid team organizations where humans and machines both pursue the same goal (Brauner et al., 2022). The example of human-robot collaboration discussed earlier stresses that communication inside the team and ergonomics in the workflow are crucial for safe and effective collaboration (Wang et al., 2017). These hybrid teams must be organized in a way that ensures acceptance by the working persons. Flexible division of tasks and mutual learning and adaption will provide methods for transferring behavior to new products. Even though the increasing degree of automation will relieve humans from having to make simple decisions, strategic decisions will still depend on humans, who will have to perceive and process increasingly complex multi-dimensional datasets and make decisions whose effects are difficult to forecast, despite all the simulation abilities built into a digital shadow. Therefore, important factors to consider include the acceptance and willingness of human actors to adopt and use novel technology, the ergonomic design of working and learning environments, and the promotion of mutual learning between humans and machines (Villani et al., 2017). Our first set of five projections in this cluster (P7, P8, P9, P10, P11) covers these thoughts.

In addition, we were interested in the effects of these new organizational structures, with regard to both acceptance and job effects. While the former aspect refers to the trust that must be generated for humans to accept and adopt the augmentation of their tasks by machines and algorithms (P12), the latter refers to the effects of the continuous automation of operational and planning processes on future employment opportunities in the production sector. While this aspect deserves an intensive investigation of its own (e.g., Sima et al., 2020), we wanted to formulate at least one projection in this domain (P13).

3.3   Capabilities

This cluster of projections deals with the capabilities that organizations need to utilize the opportunities of Next Generation Manufacturing platforms, but also the capabilities that Next Generation Manufacturing offers with regard to increasing efficiency and sustainability in manufacturing (see Table 3).

In prior research, there is a strong consensus about the capabilities required for a digital transformation of manufacturing (Warner & Wäger, 2019). This research builds on the rich literature of capability building and organizational sensemaking and studies how dedicated capabilities are linked to firm performance. With increasing amounts of production data available, the necessary qualifications of persons working with data and analytics in production processes will change significantly (Soluk and Kammerlander 2021). Intelligent decision support systems can reduce the cognitive load, but the skills necessary to interact and interpret with these systems will still be required (Brauner et al., 2022). Also, operators will have to gain the capability to managing multiple production processes or collaborate with multiple robots simultaneously (Giuliani et al., 2010; Wang et al. 2017). Another established set of capabilities is business model innovation capabilities (Burmeister et al., 2016; Bocken & Geradts, 2020), which will be required to orchestrate a manufacturing ecosystem (Kopalle et al., 2020; van Dyck et al., 2021) or enact organizational change (Björkdahl, 2020). Given the strong state of research in this area, we did not build any additional projections for these topics.

From our vision of the Internet of Production, however, we could derive a novel approach to capability building, enabled by shared digital shadows in data spaces. With such infrastructures in place, an organization could counterbalance a lack of required capabilities by having access to the abilities and skills of other actors via an

Table 3 Projections for capability configuration for and by Next Generation Manufacturing

P#	Short title	Projections for capability configuration for and by Next Generation Manufacturing
14	Expert Knowledge	In 2030, implicit expert knowledge, which traditionally could only be gained through experience, will increasingly be explicitly preserved in the form of digital models, interactive guides, or instructions and facilitated by technologies like augmented or virtual reality. As a result, this knowledge will also be made available to novices and will eliminate the dependency on experienced production employees
15	Environmental Sustainability	In 2030, environmental sustainability of production will have increased significantly compared to today
16	Production Transparency	In 2030, full transparency based on a complete digital twin of all production machines, lines, and plant engineering and a complete digital shadow of their operations will increase production efficiency significantly
17	University Degrees	In 2030, the application of biological principles (e.g., cybernetics, biomimicry) of manufacturing will have created a demand for new, multidisciplinary university degrees covering engineering, the life sciences, and computer science

IIoT platform. This is a common pattern in digital consumer markets, where a smartphone app provides consumers with access to dedicated skills (e.g., customized training plans or nutrition analysis) via a standardized platform, often building on data shared by the user's wearable device. Similar scenarios could also become possible in industrial settings. This idea of "downloading" required capabilities over a platform forms the basis of the first projection in this cluster (P14). We complemented this projection with one on the capabilities required for the ongoing biological transformation of manufacturing (Byrne et al., 2018; Neugebauer et al., 2019). While this aspect is only complementary aspect to our research, we wanted to include it through at least one projection (P17) because such a "biologicalization" of production is frequently mentioned as a future trend in manufacturing (Bergs et al., 2020; Miehe et al., 2020).

The two other projections in this section concern the capabilities enabled by a Next Generation Manufacturing system. First, the future development of industrial production and its increasing digitalization offer huge opportunities for more sustainability in Industry 4.0 (P15). The COVID-19 crisis has reinforced the focus on ecological and social sustainability (Piller et al., 2022), as argued at the beginning of this chapter. Digitalization and new value-chain constellations can lead to significant improvements in terms of lower material and energy consumption over the entire product life cycle, from engineering and production to maintenance and disposal—and society is demanding that companies utilize these opportunities (Bai et al., 2020; Ghobakhloo, 2020). But even the objective dominating today's discussion of Industry 4.0, increasing the (operational) efficiency of an established production setup, can benefit from our vision of the Internet of Production (Dalenogare et al., 2018). Hence, a final projection (P16) refers to the additional transparency in a production system that utilizes shared digital shadows to increase transparency and thus the understanding of what exactly is happening in the system and why. Transparency is a prerequisite for predictions and prescriptions.

3.4 Interfaces

This cluster of projections addresses different layers of human-machine and machine-to-machine interfaces from cognitive, spatial, and competitive perspectives (Table 4). Building on the organization dimension above, three projections cover further aspects of the design of human-machine interfaces. The increasing digitalization and intelligent connectivity of devices will lead to rising amounts of available production data, with high levels of cognitive and visual complexity required to handle these data. Although data represented in digital shadows are generally preprocessed, there are multiple application scenarios and different types of decision-makers which require context-specific visualizations of the data (Mertens et al., 2021). Decision support systems following context-sensitive design principles may provide a solution in this context by providing implicit and attentive (adaptive) support for a human operator (P18). Furthermore, advances in key technologies such as eye and body tracking, combined with greater availability of data on human movements and preferences, will allow far more implicit and natural human-machine interactions than the rather cumbersome—from an ergonomic point of view—keyboards, joysticks, and computer mice that are still predominantly used today.

Two other projections were informed by the shared experience of working from home (WFH) during the COVID-19 pandemic. There is no doubt that WFH will remain an integral part of the work organization of office-based and administrative

Table 4 Projections for interface design in Next Generation Manufacturing

P#	Short title	Projections for interface design in Next Generation Manufacturing
18	Implicit Interfaces	In 2030, human-machine interaction will have evolved away from explicit interaction, where the human operator has full control of the actions of the production system's entities, toward implicit interaction, where the system automatically adapts to the human operator's behavior by detecting and predicting their actions and modifying these actions accordingly
19	Open Interfaces	In 2030, regulatory requirements will demand open and standardized interfaces for data exchange for all kinds of manufacturing equipment
20	Production from Home	In 2030, production employees will operate their workstation from their home office, controlling, for example, remotely operated robots
21	Plant Management from Home	In 2030, plant directors will be able to manage multiple factories centrally from their home office due to the complete and real-time transparency of all the operations in a digital system

jobs. But can the possibilities of WFH be transferred to industrial production? Two projections investigate the possibility of remotely managing a plant or even operating a workstation from one's home office (P20 and P21). If these projections were to become reality, suppliers and solution providers of the corresponding digital service offerings would see a significant increase in demand. In particular, remote service offerings like virtual and augmented reality (VR/AR) technologies, autonomous robotics, and industrial metaverse applications will gain in importance. In turn, industrial work processes would be radically changed by remote services, data-based decision-making (e.g., predictive maintenance), and the increased use of digital shadows and virtual reality tools. Education and training would need to follow these developments, probably increasingly using digital delivery formats, digital learning tools, and learning analytics (Piller et al., 2020).

From the external perspective of a firm's future position on shared manufacturing (IIoT) platforms, the openness of machine-to-machine interfaces (APIs) needs special attention and became the subject of the last in this set of projections. On a technical level, the openness of APIs and other technical interfaces is not just a question of programming and quality control but is first of all an important design factor for the ability of a connected asset to provide predictive and prescriptive functionality, i.e., to enhance its capability by gaining access to data from and sending data to other actors (van Dyck et al., 2021). From the perspective of a platform, the openness of an API is a signal of a willingness to share data and knowledge and hence potentially attract third parties. At the same time, open interfaces not just are a technical risk but also reduce the ability of the originator of the data to capture unique value from this data and hence differentiate it from other market players (Adner & Kapoor, 2010). Hence, companies need to be aware of their openness decisions when designing their technical interfaces—as this will become a core strategic decision. Also, regulatory bodies may enforce larger levels of openness to prevent an uneven distribution of power toward a platform orchestrator, but also to increase the flexibility for manufacturers to switch easily between platforms and asset providers (Alexy et al., 2018; Parker & Van Alstyne, 2018). Furthermore, striving for more sustainability in industry demands the open sharing of data to build transparency along entire supply chains. Therefore, our final projection (P19) proposes that it may be regulators, and not the strategic decisions of companies, that lead to greater openness of interfaces and open data exchange between organizations.

3.5   Resilience

We complement our analytical framework with a fifth factor, resilience. In a globalized and networked economy, production interruptions, including the interruption of supply chains, have been the leading business risk for many years. But the COVID-19 pandemic demonstrated especially powerfully how unexpected events can disrupt entire global logistics chains very quickly. The resulting demand for

Table 5 Projections for drivers of future resilience in manufacturing

P#	Short title	Projections for drivers of future resilience in manufacturing
22	Decentralization	In 2030, supply chains will have become more decentralized, with production and sourcing moving closer to the end customer to cope better with global crises (e.g., pandemics)
23	Production Costs	In 2030, production costs will have increased substantially due to more regional production and higher inventory levels to cope with global crises (e.g., pandemics)
24	Production Resilience	In 2030, AI-based decision systems will enable greater resilience of production networks in the event of a global crisis (e.g., a pandemic)

change poses enormous challenges for industrial production. At the same time, the ongoing digitalization and networking of industrial value chains offer new opportunities and provide the capabilities to reach these objectives. Hence, like sustainability, resilience will complement the established set of strategic objectives of Industry 4.0, becoming similarly important to operational efficiency (OE, OEE) and strategic differentiation (e.g., individualization, flexibility, customer centricity).

The ability of a company to adapt permanently to internal and external changes and disruptions has been described as its "search for resilience" (Gu et al., 2015; Moghaddam & Deshmukh, 2019). Reinforced by a significant increase in complexity in production due to Industry 4.0, resilience management is thus becoming an indispensable success factor for production companies. But sharing data across organizational and industry boundaries and using novel predictive approaches to identify potential sources of disruption in these data, along with algorithms providing prescriptions to cope with these disruptions, promise new ways to achieve resilience. In this sense, next-generation production systems utilize data-driven ecosystems based on collaborative applications and using leading AI technologies and Industry 4.0 standards to anticipate disruptions (anticipation) and optimally adapt their production planning to mitigate against active disruptions at any time (response).

We cover these developments in a final set of projections (Table 5). Increasing resilience could become a dominant strategic objective in Next Generation Manufacturing, enabled by future applications of machine intelligence (P24). Companies could further (re-)integrate previously distributed stages of the value chain, preferring insourcing to outsourcing for a larger scope of activities. From an aggregated perspective, this would mean that supply chains would become more decentralized, with production and sourcing moving closer to customers in local markets (P22). Value chains would be shortened and more partners added for diversification, especially to secure critical components and to increase flexibility in fulfillment. For such a strategy, digitalization helps to mitigate the increasing cost of complexity resulting from vertical integration. Still, the focus on increasing resilience will overall lead to higher production cost (P23).

4   Conclusions and Outlook

In the following chapters of this study, we will present the results from the validation of our 24 projections by an international panel of experts from industry and academia and discuss the implications and insights gained from this analysis.

Chapter "Applying the Real-time Delphi Method to Next Generation Manufacturing": Marc Van Dyck, Dirk Lüttgens, and Frank T. Piller introduce the methodology of the real-time Delphi study that serves as a template for further applications of forecasting studies in interdisciplinary settings with high degrees of technical uncertainty.

Chapter "Big Picture of Next Generation Manufacturing": Sebastian Pütz, Marc Van Dyck, Dirk Lüttgens, and Alexander Mertens provide an overview of the expert assessments of all 24 projections for Next Generation Manufacturing.

Chapter "Governance Structures in Next Generation Manufacturing": Christian Brecher, Matthias Jarke, Frank T. Piller, Günther Schuh, Annika Becker, Florian Brillowski, Ester Christou, István Koren, Maximilian Kuhn, Dirk Lüttgens, Marc Van Dyck, and Marian Wiesch elaborate on how the sharing of usage data requires new forms of governance, internally and externally.

Chapter "Organization Routines in Next Generation Manufacturing": Philipp Brauner, Luisa Vervier, Florian Brillowski, Hannah Dammers, Linda Steuer-Dankert, Sebastian Schneider, Ralph Baier, Martina Ziefle, Thomas Gries, Carmen Leicht-Scholten, Alexander Mertens, and Saskia K. Nagel discuss how the introduction of digital shadows will impact the organization of work, workers, and workplace design.

Chapter "Capability Configuration in Next Generation Manufacturing": Christian Hinke, Luisa Vervier, Philipp Brauner, Sebastian Schneider, Linda Steuer-Dankert, Martina Ziefle, and Carmen Leicht-Scholten show how the digital transformation affects the capabilities of production systems and companies, as well as requirements for higher education and educational programs.

Chapter "Interface Design in Next Generation Manufacturing": Ralph Baier, Srikanth Nouduri, Luisa Vervier, Philipp Brauner, István Koren, Martina Ziefle, and Verena Nitsch discuss the future of emerging trends in human-machine interaction such as implicit interfaces and teleoperation from home.

Chapter "Resilience Drivers in Next Generation Manufacturing": Alexander Schollemann, Marian Wiesch, Christian Brecher, and Günther Schuh illustrate opportunities for improving the resilience of global production networks via decentralization and the use of AI-based decision support systems.

Chapter "Future Scenarios and the Most Probable Future for Next Generation Manufacturing": Marc Van Dyck, Sebastian Pütz, Alexander Mertens, Dirk Lüttgens, Verena Nitsch, and Frank T. Piller present scenarios that portray the most probable future for Next Generation Manufacturing in 2030.

We close this book with a chapter discussing a core pattern that we see in many of the implications presented in the previous chapters. Today, humans and machines/algorithms mostly work sequentially with each other: an algorithm provides decision support, but the human makes the final decision. Or, vice versa, humans generate or engage in pre-classification of data, and an autonomous system then makes and executes a decision. But the results from our Delphi study predict that humans and machines will become real collaborative partners in the near future—a development we call "hybrid intelligence" (following Dellermann et al., 2019). For manual work, the rise of collaborative robots has already begun this shift. But for the planning and engineering levels, this change is still to come. Hence, chapter "Hybrid Intelligence in Next Generation Manufacturing: An Outlook on New Forms of Collaboration between Human and Algorithmic Decision Makers in the Factory of the Future" introduces the concept of hybrid intelligence, complementing the results from the Delphi study and providing a further outlook on the future of manufacturing.

Acknowledgment Funded by the Deutsche Forschungsgemeinschaft (DFG, German Research Foundation) under Germany's Excellence Strategy—EXC-2023 Internet of Production—390621612.

References

Adner, R. (2017). Ecosystem as structure: An actionable construct for strategy. *Journal of Management, 43*(1), 39–58. https://doi.org/gc8svt

Adner, R., & Kapoor, R. (2010). Value creation in innovation ecosystems: How the structure of technological interdependence affects firm performance in new technology generations. *Strategic Management Journal, 31*(3), 306–333. https://doi.org/b2wksf

Agrawal, M., Eloot, K., Mancini, M., & Patel, A. (2020). Industry 4.0: Reimagining manufacturing operations after COVID-19. McKinsey whitepaper. July 2020.

Ahlers, E. (2016). Flexible and remote work in the context of digitization and occupational health. *International Journal of Labour Research, 8*(1–2), 85–99.

Alexy, O., West, J., Klapper, H., & Reitzig, M. (2018). Surrendering control to gain advantage: Reconciling openness and the resource-based view of the firm. *Strategic Management Journal, 39*(6), 1704–1727. https://doi.org/gcp2zz

Bai, C., Dallasega, P., Orzes, G., & Sarkis, J. (2020). Industry 4.0 technologies assessment: A sustainability perspective. *International Journal of Production Economics, 229*, 107776. https://doi.org/gg7mf2

Bauernhansl, T., Hartleif, S., & Felix, T. (2018). The digital shadow of production: A concept for the effective and efficient information supply in dynamic industrial environments. *Procedia CIRP, 72*, 69–74. https://doi.org/gfgvgs

Becker, F., Bibow, P., Dalibor, M., Jarke, M., et al. (2021). A conceptual model for digital shadows in industry and its application. In *Proceedings of the international conference on conceptual modeling* (pp. 271–281). Springer. https://doi.org/hg45

Bergs, T., Schwaneberg, U., Barth, S., Hermann, L., Grunwald, T., Mayer, S., ... Sözer, N. (2020). Application cases of biological transformation in manufacturing technology. *CIRP Journal of Manufacturing Science and Technology, 31*, 68–77. https://doi.org/hhn7

Björkdahl, J. (2020). Strategies for digitalization in manufacturing firms. *California Management Review, 62*(4), 17–36. https://doi.org/ggv59w

Bocken, N. M., & Geradts, T. H. (2020). Barriers and drivers to sustainable business model innovation: Organization design and dynamic capabilities. *Long Range Planning, 53*(4), 101950. https://doi.org/ghd4fb

Boschert, S., Heinrich, C., & Rosen, R. (2018). Next generation digital twin. In *Proceedings of TMCE* (pp. 209–218) Las Palmas de Gran Canaria, Spain.

Brauner, P., Dalibor, M., Jarke, M., Kunze, I., Koren, I., Lakemeyer, G., … Ziefle, M. (2022). A computer science perspective on digital transformation in production. *ACM Transactions on Internet of Things, 3*(2), 1–32. https://doi.org/10.1145/3502265

Brecher, C., Özdemir, D., & Weber, A. R. (2016). Integrative production technology—Theory and applications. In C. Brecher & D. Özdemir (Eds.), *Integrative production technology* (pp. 1–17). Springer. https://doi.org/hhn9

Burmeister, C., Lüttgens, D., & Piller, F. T. (2016). Business model innovation for Industrie 4.0: Why the industrial internet mandates a new perspective on innovation. *Die Unternehmung, 70*(2), 125–140. https://doi.org/hhpb

Byrne, G., Dimitrov, D., Monostori, L., Teti, R., van Houten, F., & Wertheim, R. (2018). Biologicalisation: Biological transformation in manufacturing. *CIRP Journal of Manufacturing Science and Technology, 21*, 1–32. https://doi.org/hhpc

Cappiello, C., Gal, A., Jarke, M., & Rehof, J. (2020). Data ecosystems: Sovereign data exchange among organizations (Dagstuhl seminar 19391). *Dagstuhl Reports, 9*(9), 66–134. https://doi.org/hhk5

Charles, R. L., & Nixon, J. (2019). Measuring mental workload using physiological measures: A systematic review. *Applied Ergonomics, 74*, 221–232. https://doi.org/gf7tds

Dahlander, L., Gann, D. M., & Wallin, M. W. (2021). How open is innovation? A retrospective and ideas forward. *Research Policy, 50*(4), 104218. https://doi.org/gjg9d4

Dalenogare, L. S., Benitez, G. B., Ayala, N. F., & Frank, A. G. (2018). The expected contribution of industry 4.0 technologies for industrial performance. *International Journal of Production Economics, 204*, 383–394. https://doi.org/gff78f

Dattée, B., Alexy, O., & Autio, E. (2018). Maneuvering in poor visibility: How firms play the ecosystem game when uncertainty is high. *Academy of Management Journal, 61*(2), 466–498. https://doi.org/gdsh4z

Dellermann, D., Ebel, P., Söllner, M., & Leimeister, J. M. (2019). Hybrid intelligence. *Business & Information Systems Engineering, 61*(5), 637–643. https://doi.org/ggkxz4

Drehborg, K. H. (1996). Essence of Backcasting. *Futures, 28*(9), 813–828. https://doi.org/bdn98t

ElMaraghy, H., Monostori, L., Schuh, G., & ElMaraghy, W. (2021). Evolution and future of manufacturing systems. *CIRP Annals, 70*(2), 635–658. https://doi.org/gn4t4r

Fleisch, E., Weinberger, M., & Wortmann, F. (2014). Geschäftsmodelle im Internet der Dinge. *HMD Praxis der Wirtschaftsinformatik, 51*(6), 812–826. https://doi.org/ggsdn4

Gawer, A. (2014). Bridging differing perspectives on technological platforms: Toward an integrative framework. *Research Policy, 43*(7), 1239–1249. https://doi.org/gc8sc5

Ghobakhloo, M. (2020). Industry 4.0, digitization, and opportunities for sustainability. *Journal of Cleaner Production, 252*, 119869. https://doi.org/gg44vk

Giuliani, M., Lenz, C., Müller, T., Rickert, M., & Knoll, A. (2010). Design principles for safety in human-robot interaction. *International Journal of Social Robotics, 2*(3), 253–274. https://doi.org/c7sf7h

Gnatzy, T., Warth, J., von der Gracht, H., & Darkow, I. L. (2011). Validating an innovative real-time Delphi approach-a methodological comparison between real-time and conventional Delphi studies. *Technological Forecasting and Social Change, 78*(9), 1681–1694. https://doi.org/ddjfk5

Gordon, T., & Pease, A. (2006). RT Delphi: An efficient, "round-less" almost real time Delphi method. *Technological Forecasting and Social Change, 73*(4), 321–333. https://doi.org/b7q893

Gries, T. (2020). Survival through innovation - understanding COVID-19 as an opportunity rather than a threat. In *Proceedings of the International Symposium of the RRI Robot Revolution Imitative. Joint Conference by Engineering Academy Japan / Acatech. November 2011.*

Gubbi, J., Buyya, R., Marusic, S., & Palaniswami, M. (2013). Internet of Things (IoT): A vision architectural elements and future directions. *Future Generation Computer Systems, 29*(7), 1645–1660. https://doi.org/10.1016/j.future.2013.01.010

Gu, X., Jin, X., Ni, J., & Koren, Y. (2015). Manufacturing system design for resilience. *Procedia Cirp, 36*, 135–140. https://doi.org/hhpd

Guth, J., Breitenbücher, U., Falkenthal, M., Leymann, F., & Reinfurt, L. (2016). Comparison of IoT platform architectures: A field study based on a reference architecture. In 2016 Cloudification of the Internet of Things (CioT) (pp. 1-6). https://doi.org/ggc8hp

Hartmann, D., & Van der Auweraer, H. (2021). Digital twins. In M. Cruz, C. Parés, & P. Quintela (Eds.), *Progress in Industrial Mathematics: Success Stories* (SEMA SIMAI Springer Series) (Vol. 5). Springer. https://doi.org/hhpf

Hedenstierna, C. P. T., Disney, S. M., Eyers, D. R., Holmström, J., Syntetos, A. A., & Wang, X. (2019). Economies of collaboration in build-to-model operations. *Journal of Operations Management, 65*(8), 753–773. https://doi.org/ghqwh8

Hirsch-Kreinsen, H., & Ittermann, P. (2021). Digitalization of work processes: A framework for human-oriented work design. In *The Palgrave handbook of workplace innovation* (pp. 273–293). Palgrave Macmillan. https://doi.org/hhpg

Hoffmann, J. B., Heimes, P., & Senel, S. (2018). IoT platforms for the internet of production. *IEEE Internet of Things Journal, 6*(3), 4098–4105. https://doi.org/ggqvpc

Holmberg, J., & Robèrt, K. (2000). Backcasting: A framework for strategic planning. *International Journal of Sustainable Development & World Ecology, 7*(4), 291–308. https://doi.org/dc7t55

Iansiti, M., & Lakhani, K. (2020). Competing in the age of AI. *Harvard Business Review, 98*(1), 59–67.

Kopalle, P. K., Kumar, V., & Subramaniam, M. (2020). How legacy firms can embrace the digital ecosystem via digital customer orientation. *Journal of the Academy of Marketing Science, 48*(1), 114–131. https://doi.org/gj27r5

Kortmann, S., & Piller, F. (2016). Open business models and closed-loop value chains: Redefining the firm-consumer relationship. *California Management Review, 58*(3), 88–108. https://doi.org/gf82bc

Liebenberg, M., & Jarke, M. (2020). Information systems engineering with digital shadows: Concept and case studies. In Proceedings of the 32nd International Conference on Advanced Information Systems Engineering (Vol. 12127, pp. 70–84). Springer. https://doi.org/10.1007/978-3-030-49435-3_5

Mertens, A., Pütz, S., Brauner, P., Brillowski, F., Buczak, N., Dammers, H., . . . Nitsch, V. (2021). Human Digital Shadow: Data-based Modeling of Users and Usage in the Internet of Production. In *2021 14th International Conference on Human System Interaction (HSI)* (pp. 1–8) https://doi.org/hg6g

Miehe, R., Bauernhansl, T., Beckett, M., Brecher, C., Demmer, A., Drossel, W. G., . . . Wolperdinger, M. (2020). The biological transformation of industrial manufacturing–technologies, status and scenarios for a sustainable future of the German manufacturing industry. *Journal of Manufacturing Systems, 54*, 50–61. https://doi.org/hhpj

Moghaddam, M., & Deshmukh, A. (2019). Resilience of cyber-physical manufacturing control systems. *Manufacturing Letters, 20*, 40–44. https://doi.org/hhpk

Mütze-Niewöhner, S., Mayer, C., Harlacher, M., Steireif, N., & Nitsch, V. (2022). Work 4.0: Human-centered work Design in the Digital age. In W. Frenz (Ed.), *Handbook industry 4.0: Law, technology, society.* Springer.

Nelles, J., Kuz, S., Mertens, A., & Schlick, C. M. (2016). Human-centered design of assistance systems for production planning and control: The role of the human in industry 4.0. In *2016 IEEE International Conference on Industrial Technology (ICIT)* (pp. 2099–2104) https://doi.org/hhpm

Neugebauer, R., Ihlenfeldt, S., Schliessmann, U., Hellmich, A., & Noack, M. (2019). A new generation of production with cyber-physical systems enabling the biological transformation in manufacturing. *Journal of Machine Engineering, 19*(1), 5–15. https://doi.org/hhpn

Otto, B., Hompel, M., & Wrobel, S. (2019). International data spaces. In R. Neugebauer (Ed.), *Digital transformation.* Springer. https://doi.org/hhpp

Otto, B., & Jarke, M. (2019). Designing a multi-sided data platform: Findings from the international data spaces case. *Electronic Markets, 29*(4), 561–580. https://doi.org/ggqvq9

Parasuraman, R., Sheridan, T. B., & Wickens, C. D. (2000). A model for types and levels of human interaction with automation. IEEE transactions on systems, man, and cybernetics. *Part A: Systems and Humans, 30*(3), 286–297. https://doi.org/c6zf92

Parker, G. G., & Van Alstyne, M. W. (2018). Innovation, openness, and platform control. *Management Science, 64*(7), 3015–3032.

Piller, F., Falk, S. et al. (2020). Ten propositions on the future of digital business models for Industry 4.0 in the post-corona economy. Position paper of the German Stakeholder Platform Industrie 4.0. Berlin: 2020. https://doi.org/hhpq

Piller, F. T, et al. (2022). Industry 4.0 and sustainability: How digital business models foster sustainability in industry. Position paper of the German Stakeholder Platform Industrie 4.0. Berlin: 2020. https://doi.org/hhk6

Porter, M. E., & Heppelmann, J. E. (2015). How smart, connected products are transforming companies. *Harvard Business Review, 93*(10), 96–114.

Reischauer, G. (2018). Industry 4.0 as policy-driven discourse to institutionalize innovation systems in manufacturing. *Technological Forecasting and Social Change, 132*, 26–33. https://doi.org/ggwnfj

Riesener, M., Schuh, G., Dölle, C., & Tönnes, C. (2019). The digital shadow as enabler for data analytics in product life cycle management. *Procedia CIRP, 80*, 729–734. https://doi.org/ggzs5d

Rochet, J. C., & Tirole, J. (2003). Platform competition in two-sided markets. *Journal of the European Economic Association, 1*(4), 990–1029. https://doi.org/b45krd

Schlick, C., Bruder, R., & Luczak, H. (2018). *Arbeitswissenschaft*. Springer.

Schuh, G., Gützlaff, A., Sauermann, F., & Maibaum, J. (2020). Digital shadows as an enabler for the internet of production. In *IFIP International Conference on Advances in Production Management Systems* (pp. 179–186). Springer. https://doi.org/hhps

Shin, D. (2014). A socio-technical framework for internet-of-things design: A human-centered design for the internet of things. *Telematics and Informatics, 31*(4), 519–531. https://doi.org/hhpt

Sima, V., Gheorghe, I. G., Subić, J., & Nancu, D. (2020). Influences of the industry 4.0 revolution on the human capital development and consumer behavior. *Sustainability, 12*(10), 4035. https://doi.org/gmg82p

Sisinni, E., Saifullah, A., Han, S., Jennehag, U., & Gidlund, M. (2018). Industrial internet of things: Challenges, opportunities, and directions. *IEEE Transactions on Industrial Informatics, 14*(11), 4724–4734. https://doi.org/gd8t5n

Sjödin, D. R., Parida, V., & Wincent, J. (2016). Value co-creation process of integrated product-services: Effect of role ambiguities and relational coping strategies. *Industrial Marketing Management, 56*, 108–119. https://doi.org/f3pt8q

Snihur, Y., Zott, C., & Amit, R. (2021). Managing the value appropriation dilemma in business model innovation. *Strategy Science, 6*(1), 22–38. https://doi.org/gnz49v

Soluk, J., & Kammerlander, N. (2021). Digital transformation in family-owned Mittelstand firms: A dynamic capabilities perspective. *European Journal of Information Systems, 30*(6), 676–711. https://doi.org/gjf66b

Taleb, N. (2005). *The black swan: Why don't we learn that we don't learn*. Random House.

Teece, D. J., Raspin, P. G., & Cox, D. R. (2020). Plotting strategy in a dynamic world. *MIT Sloan Management Review, 62*(1), 28–33.

Thorade, N. (2020). *Vernetzte Produktion: Computer Integrated Manufacturing (CIM) als Vorgeschichte von Industrie 4.0*. Friedrich-Ebert-Stiftung.

Van Dyck, M., Lüttgens, D., Piller, F., & Diener, K. (2021). Positioning strategies in emerging industrial ecosystems for industry 4.0. In *Proceedings of the 54th Hawaii International Conference on System Sciences, January 2021* (pp. 6153–6162).

Villani, V., Sabattini, L., Czerniaki, J. N., Mertens, A., Vogel-Heuser, B., & Fantuzzi, C. (2017). Towards modern inclusive factories: A methodology for the development of smart adaptive human-machine interfaces. In *2017 22nd IEEE international conference on emerging technologies and factory automation (ETFA)* (pp. 1–7) https://doi.org/hhpv

Wang, X. V., Kemény, Z., Váncza, J., & Wang, L. (2017). Human–robot collaborative assembly in cyber-physical production: Classification framework and implementation. *CIRP Annals, 66*(1), 5–8. https://doi.org/gbth55

Warner, K. S., & Wäger, M. (2019). Building dynamic capabilities for digital transformation: An ongoing process of strategic renewal. *Long Range Planning, 52*(3), 326–349. https://doi.org/gf6prh

Wilkesmann, U. (2005). Die Organisation von Wissensarbeit. *Berliner Journal für Soziologie, 15*(1), 55–72. https://doi.org/br7g6v

Applying the Real-Time Delphi Method to Next Generation Manufacturing

Marc Van Dyck, Dirk Lüttgens, and Frank T. Piller

Abstract The Delphi method is a structured scientific approach used to organize and structure an expert discussion in order to gain insights about the future. In order to develop scenarios for the future of Next Generation Manufacturing, an innovative real-time Delphi survey was conducted with 35 experts from industry and academia. The survey involved evaluating a set of 24 projections on the future of Next Generation Manufacturing, and the results of the survey were used to develop reliable future scenarios. Our main objective was to create a picture of the elements of Next Generation Manufacturing in 2030, guided by developments in the context of Industry 4.0. By using an innovative real-time Delphi approach in the context of Next Generation Manufacturing, we extend this established tool of strategic technology management from predicting technological developments and their impact on firms and society to providing a strategic guide for decision-makers in times of high uncertainty. Our study thus serves as a template for further applications of forecasting studies in interdisciplinary settings with high degrees of technical uncertainty.

[Abstract generated by machine intelligence with GPT-3. No human intelligence applied.]

M. Van Dyck (✉) · D. Lüttgens
Institute for Technology and Innovation Management, RWTH Aachen University, Aachen, Germany
e-mail: vandyck@time.rwth-aachen.de; luettgens@time.rwth-aachen.de

F. T. Piller
Institute for Technology and Innovation Management, RWTH Aachen University, Aachen, Germany

Institute for Business Cybernetics (IfU) e.V. at RWTH Aachen, Aachen, Germany
e-mail: piller@time.rwth-aachen.de

© The Author(s), under exclusive license to Springer Nature Switzerland AG 2022
F. T. Piller et al. (eds.), *Forecasting Next Generation Manufacturing*, Contributions to Management Science, https://doi.org/10.1007/978-3-031-07734-0_2

1   Scenario Development for Next Generation Manufacturing

Technological developments can have an impact on firms and society. This impact is often unpredictable, creating a need to manage the involved uncertainty (IPTS Economists Group et al., 2002). A common approach to managing uncertainty is engaging in forecasting projects which involve the generation of future scenarios that outline influencing factors and trends (Gausemeier et al., 1998). The main goal of such forecasting efforts is to anticipate the future (Saritas & Oner, 2004) and to serve as a basis for long-term planning (Courtney et al., 1997).

Similarly, the implications for Next Generation Manufacturing are unclear given the high uncertainty of technological development involved. We provide an approach to developing scenarios for future outcomes of Next Generation Manufacturing which can enable strategic planning by firms and future research. Our research is guided by one core question: *how will digital shadows influence manufacturing firms from the perspective of employees, managers, firms, and society?*

Scenario development is done using technological forecasting methods, which have a long tradition in strategic technology management. One can distinguish three categories of forecasting methods. First, *exploratory methods* project current technological progress into the future, for example, trend exploration or bibliometric analyses (Cho & Daim, 2013). Second, *normative methods*, such as multi-criteria decision models or morphological analyses, illustrate the path to a desired future (Roberts, 1969). Third, *combined methods* integrate both approaches, such as the Delphi method (TFAMW Group, 2004). As our aim is to assess the probability and impact of early-stage technology on a diverse set of stakeholder groups, we followed common practice and drew on the Delphi method, an expert-based assessment (Landeta, 2006).

The Delphi method is a structured scientific approach to organizing and structuring an expert discussion in order to gain insights (Beiderbeck et al., 2021). Its purpose is to derive a reliable consensus about future developments by structuring complex opinions from various stakeholders (Kameoka et al., 2004; Linstone & Turoff, 2002; Rauch, 1979). It is considered a "judgmental forecasting procedure" (von der Gracht & Darkow, 2010), is constructed in an interactive multi-stage format, and is conducted anonymously and in written form. Here, experts assess statements about the future, so-called projections. Given the complexity of the problems, it is crucial to incorporate diverse perspectives in terms of both the set of projections and the selection of experts (Linstone, 1981). In addition, Saritas and Oner (2004) suggest including comments by the experts explaining the reasoning for their quantitative estimates.

Our scenario development is built on a multi-round, real-time Delphi survey with 35 experts from industry and academia who evaluate a set of 24 projections. In the following, we outline our process for conducting the Delphi survey. We applied a platform framework, adapted from Gawer (2014), distinguishing four dimensions: governance (e.g., open forms of collaboration), organization (e.g., boundaries and decision-making), capabilities (e.g., hybrid intelligence), and interfaces (e.g., open APIs and human-machine interfaces). In addition, we added a section on resilience drivers to framework for the development of the projections and the scenarios. Conventional Delphi surveys face criticism regarding a failure to translate their findings into actionable results due to being a time-consuming process (Gnatzy et al., 2011) with high drop-out rates (Keller & von der Gracht, 2014). Thus, we used a novel real-time Delphi approach, as described by Gordon and Pease (2006) and improved by Gnatzy et al. (2011). In this approach, experts evaluate the projections through an interactive online interface that provides instant feedback in the form of the other experts' assessments and allows the participants to engage in discussion and potentially to adjust their estimations. As well as ensuring anonymity, the internet-based approach is more efficient and accessible, thus reducing drop-out rates and increasing the accuracy of the results (Gnatzy et al., 2011). A sample real-time Delphi survey can be found in Jiang et al. (2017).

2 Real-Time Delphi Process

Strict adherence to a rigorous process is key to ensuring the reliability and validity of a Delphi survey (Hasson & Keeney, 2011). We followed a four-step process to deliver this, as suggested by von der Gracht and Darkow (2010): first, we developed our Delphi projections; second, we selected a panel of experts; third, we conducted the Delphi survey; and finally, we developed future scenarios (see Fig. 1). We will provide a detailed overview and discuss the results of each step in the subsequent sections.

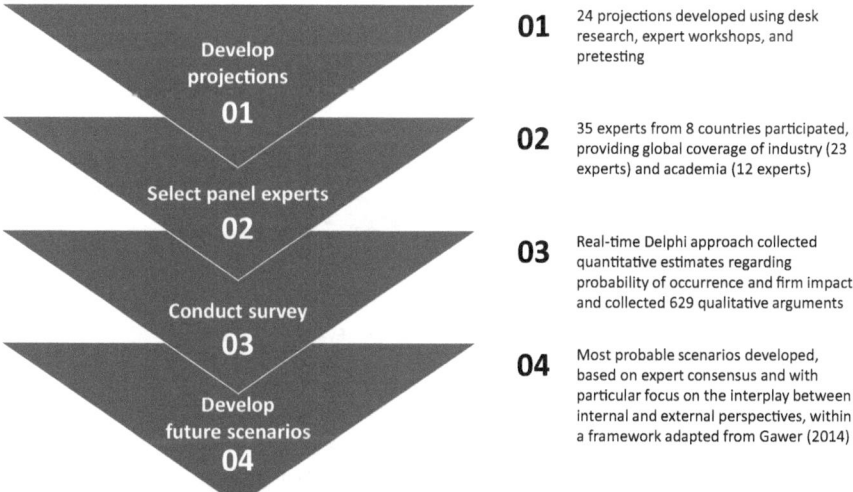

Fig. 1 Real-time Delphi process

2.1   Step 1: Develop Delphi Projections

First, we developed a set of projections for Next Generation Manufacturing. In line with previous Delphi studies (e.g., Jiang et al., 2017; von der Gracht & Darkow, 2010), we chose 2030 as the projection horizon for our scenarios, giving a 10-year timeframe. To address the required diversity of perspectives, we used an adaptation of Gawer's (2014) framework to structure the formulation of our projections, as outlined previously. We conducted workshops with 27 experts from the fields of computer science, engineering, management, and social sciences to develop projections. Workshop participants did not participate in the survey. In addition, we used literature research to triangulate the workshop results (Gausemeier et al., 1998). As a result, we identified an initial set of 76 projections. We clustered similar projections to rule out redundancy and ensure an equal level of detail. Hence, we reduced the number of projections to 45. To ensure that we gained valid results without causing research fatigue and to guarantee that we covered all relevant topics within our framework dimensions, we went back to our workshop participants to discuss the reduced set of projections. After this second evaluation, we were able to dramatically reduce the number of projections again, to 24. In addition to the number of projections, we paid special attention to their quality and comprehensibility. Short, unequivocal, and precise wording is key to avoid any ambiguity which would impact the quality of the outcome (Mićić, 2007). Figure 2 illustrates the process of developing the projections.

We conducted a pre-test with 13 experts from industry and research to ensure content reliability as well as face validity. The pre-tested set of 24 projections then underwent a final editing round before being presented to the panel experts using the internet-based real-time Delphi tool developed by Gnatzy et al. (2011).

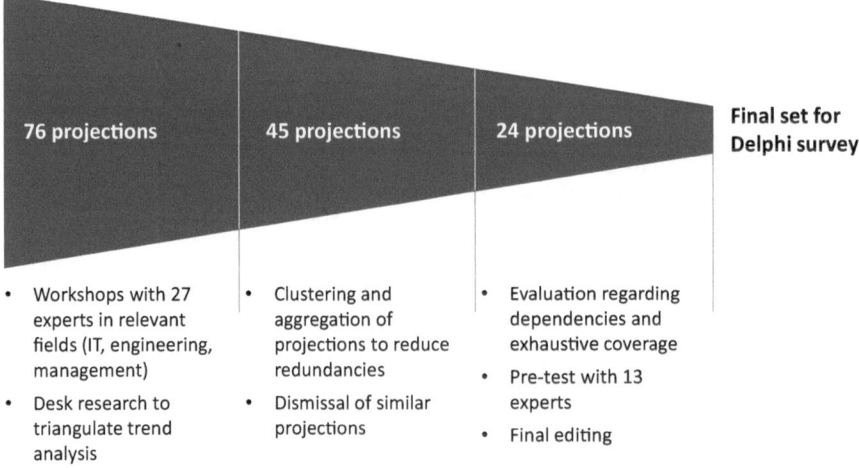

Fig. 2 Projection development funnel

2.2   Step 2: Select Panel Experts

In a second step, we composed a panel of experts by identifying, selecting, and recruiting relevant actors in the field of manufacturing, in particular digital manufacturing (Gordon & Pease, 2006). Panel sizes in Delphi studies vary and depend on the scope of the study, required heterogeneity, and availability (Loo, 2002). In previous Delphi studies, the panel size has ranged from 10 to 60 participants (e.g., Beiderbeck et al., 2021; Gordon & Helmer, 1964; Jiang et al., 2017). Our target panel size was in the middle of this range, as digital shadows involve a heterogeneous actor structure but are still an emerging field with a limited number of available experts.

We identified potential experts by tapping into the network of all the workshop participants, as well as by searching in professional social networks such as LinkedIn. We reached out to this initial set of experts and asked them to refer experts who are more knowledgeable than they themselves, following an approach known as a pyramiding search (von Hippel et al., 2009). Selection criteria included technical expertise, publications in the relevant field, and profession. We evaluated the experts according to their corporate function, company's stake in the technology, or previous publications and adapted the initial set to include a diverse group of experts. Our final panel contained 35 experts, including different stakeholders from industry (23) and academia (12) across a global range of nationalities (8). Table 1 provides an overview of our final expert panel. To our knowledge, our panel is one of the most comprehensive used in a study on digital shadows.

2.3   Step 3: Conduct Survey

For the Delphi survey, we used the real-time survey software developed by Gnatzy et al. (2011). The participating experts were presented with one projection at a time to reduce information overload. First, we asked the experts to provide their estimates on the probability of occurrence and the impact on firms of the projection in the year 2030. Probability of occurrence was measured in percent (0–100%), while firm impact was measured on a 5-point Likert scale (1 = very low to 5 = very high). In addition, we invited the experts to provide qualitative comments explaining the reasoning for their estimates. We were able to collect a large amount of qualitative data, with 629 comments. This indicates the commitment of the participants as well as their relevant expertise.

After the experts had provided their initial estimates and reasoning for a given projection, the next page presented the intermediate results (mean, standard deviation, interquartile range), as well as the anonymized arguments of the other experts for this projection. In line with the aim of the Delphi method to reach a consensus, the experts were prompted to reevaluate their estimated probability of occurrence and firm impact. In addition, they could engage in a discussion anonymously by

Table 1 Expert panel

#	Affiliation	Region	Field	Competency
1	Academia	Europe	Information systems	Professor for computational analysis of technical systems
2	Industry	Europe	Aerospace	Digital transformation manager
3	Industry	Europe	Automotive	Data scientist
4	Industry	Europe	Consulting	Consultant in industrial complexity management
5	Industry	Europe	Conglomerate	R&D strategy consultant
6	Academia	North America	Engineering	Professor for manufacturing systems
7	Industry	Europe	Industrial equipment	Expert in manufacturing excellence
8	Industry	Europe	Industrial equipment	Executive vice president
9	Academia	North America	Engineering	Professor of mechanical and aerospace engineering
10	Academia	Europe	Engineering	Professor of prognostics and health management
11	Academia	Europe	Engineering	Senior researcher for applied industrial engineering and ergonomics
12	Industry	Europe	Conglomerate	Expert in additive manufacturing
13	Industry	Europe	Aerospace	Director of production
14	Industry	Europe	Consulting	Managing director and partner, global leader in manufacturing
15	Academia	Europe	Engineering	Professor of production systems
16	Industry	Asia	Electronics	Vice chairman and board member
17	Industry	Europe	Industrial software	Managing director
18	Academia	North America	Engineering	Professor of manufacturing engineering
19	Academia	Europe	Economics	Professor of economics and entrepreneurship
20	Industry	Europe	Automotive	Industrial engineer
21	Industry	Europe	Chemicals	Innovation manager
22	Industry	Asia	Conglomerate	Senior chief researcher
23	Academia	Europe	Information systems	Professor of software and systems engineering
24	Industry	Europe	Industrial equipment	Head of product marketing
25	Academia	Europe	Information systems	Professor of business informatics and data science
26	Industry	Europe	Industrial software	Chief technology officer
27	Industry	Europe	Automotive	Director of manufacturing
28	Industry	Europe	Industrial equipment	Managing director

(continued)

Table 1 (continued)

#	Affiliation	Region	Field	Competency
29	Industry	Europe	Industrial equipment	R&D manager, laser technology
30	Industry	Europe	Textile manufacturer	Head of finance
31	Academia	Europe	Engineering	Professor of production planning and control
32	Industry	Europe	Aerospace	Founder and technical director for lightweight construction parts
33	Industry	Europe	Materials	Chief technology officer
34	Industry	Europe	Automotive	Head of operations, production support
35	Academia	North America	Economics	Professor of management

responding to other experts' comments. Thereby, we strengthened the data's validity (von der Gracht & Darkow, 2010).

2.4   Step 4: Develop Future Scenarios

In a final step, we used the Delphi results to derive future scenarios regarding the probability of occurrence and impact on firms of digital shadows within Next Generation Manufacturing in 2030. For this, we first analyzed the quantitative results of the Delphi survey by calculating the mean, standard deviation, interquartile range, and outliers for each projection. To identify whether a consensus was reached among our expert panel for a projection, we used the interquartile range, which measures the difference between the upper and lower quartiles (Sekaran & Bougie, 2013). In line with previous Delphi studies (e.g., Jiang et al., 2017; von der Gracht & Darkow, 2010), we considered each projection with an interquartile range equal to or less than 2.0, indicating a low dispersion from the median, as having a consensus. Consensus was measured for the probability of occurrence as we developed our future scenarios using the most probable scenarios, even if there was a higher dispersion among the firm impact values.

After establishing the quantitative baseline for each projection, we described the results for each projection separately. In addition to the quantitative estimates, we analyzed the qualitative comments. They provided a richer understanding of and reasoning with which to interpret the quantitative estimates. For this, we took the results back to the initial workshop group that had developed the projections in the first place. Workshop participants clustered around projections they were particularly knowledgeable about. In this way, we were not only able to reflect a diversity of perspectives in the survey, but we were also able to incorporate heterogeneous perspectives in our interpretation of the results by including interdisciplinary research teams in the interpretation process. The quantitative results provided the basis for our analysis. The qualitative comments were coded and aggregated to broader themes and served as complementary data. We then followed a clear structure, describing the results for each projection, providing use cases from industry and academia, and outlining implications for policy makers, firms, and individuals (managers, employees).

Finally, we developed future scenarios by clustering selected projections, as suggested by von der Gracht and Darkow (2010). We followed the previously introduced framework adopted from Gawer (2014) and developed the most probable scenarios for each dimension according to the aggregated statistics from the survey (for the aggregated statistics, see in chapter "Big Picture of Next Generation Manufacturing"; and for a synthesis of the results, see chapter "Governance Structures in Next Generation Manufacturing").

3 Summary

Managing the uncertainty resulting from technological developments is paramount to prepare for potential scenarios. By using an innovative real-time Delphi approach in the context of Next Generation Manufacturing, we extend this established tool of strategic technology management from predicting technological developments and their impact on firms and society (Courtney et al., 1997) to a providing a strategic guide for decision-makers in times of high uncertainty. Our study thus serves as a template for further applications of forecasting studies in interdisciplinary settings with high degrees of technical uncertainty.

Acknowledgment Funded by the Deutsche Forschungsgemeinschaft (DFG, German Research Foundation) under Germany's Excellence Strategy – EXC-2023 Internet of Production – 390621612.

References

Beiderbeck, D., Frevel, N., Heiko, A., Schmidt, S. L., & Schweitzer, V. M. (2021). Preparing, conducting, and analyzing Delphi surveys: Cross-disciplinary practices, new directions, and advancements. *MethodsX, 8*, 101401. https://doi.org/gmpjvv

Cho, Y., & Daim, T. (2013). Technology forecasting methods. In T. U. Daim, T. Oliver, & J. Kim (Eds.), *Research and technology management in the electricity industry: Methods, tools and case studies* (pp. 67–112). Springer. https://doi.org/hg4q

Courtney, H., Kirkland, J., & Viguerie, P. (1997). Strategy under uncertainty. *Harvard Business Review, 75*(6), 67–79.

Gausemeier, J., Fink, A., & Schlake, O. (1998). Scenario management: An approach to develop future potentials. *Technological Forecasting and Social Change, 59*(2), 111–130. https://doi.org/dz753z

Gawer, A. (2014). Bridging differing perspectives on technological platforms: Toward an integrative framework. *Research Policy, 43*(7), 1239–1249. https://doi.org/gc8sc5

Gnatzy, T., Warth, J., von der Gracht, H., & Darkow, I. L. (2011). Validating an innovative real-time Delphi approach-a methodological comparison between real-time and conventional Delphi studies. *Technological Forecasting and Social Change, 78*(9), 1681–1694. https://doi.org/ddjfk5

Gordon, T., & Pease, A. (2006). RT Delphi: An efficient, "round-less" almost real time Delphi method. *Technological Forecasting and Social Change, 73*(4), 321–333. https://doi.org/b7q893

Gordon, T. J., & Helmer, O. (1964). *Report on a long-range forecasting study*. Rand Corp.

Hasson, F., & Keeney, S. (2011). Enhancing rigour in the Delphi technique research. *Technological Forecasting and Social Change, 78*(9), 1695–1704. https://doi.org/dzd92c

IPTS Economists Group, Branson, W., Laffont, J. J., Solow, R., Ulph, D., von Weizsäcker, C., & Kyriakou, D. (2002). Economic dimensions of prospective technological studies at the Joint Research Centre of the European Commission. *Technological Forecasting and Social Change, 69*(9), 851–859. https://doi.org/c29x6w

Jiang, R., Kleer, R., & Piller, F. T. (2017). Predicting the future of additive manufacturing: A Delphi study on economic and societal implications of 3D printing for 2030. *Technological Forecasting and Social Change, 117*, 84–97. https://doi.org/ghzgpp

Kameoka, A., Yokoo, Y., & Kuwahara, T. (2004). A challenge of integrating technology foresight and assessment in industrial strategy development and policymaking. *Technological Forecasting and Social Change, 71*(6), 579–598. https://doi.org/d7twt4

Keller, J., & von der Gracht, H. A. (2014). ICT and the foresight infrastructure of the future: Effects on the foresight discipline. *World Future Review, 6*(1), 40–47. https://doi.org/hg4r

Landeta, J. (2006). Current validity of the Delphi method in social sciences. *Technological Forecasting and Social Change, 73*(5), 467–482. https://doi.org/ckpjp5

Linstone, H. A. (1981). The multiple perspective concept. *Technological Forecasting and Social Change, 20*(4), 275–325. https://doi.org/b7h9sb

Linstone, H. A., & Turoff, M. (2002). *The Delphi method: Techniques and applications*. Addison-Wesley.

Loo, R. (2002). The Delphi method: A powerful tool for strategic management. *Policing: An International Journal of Police Strategies & Management, 25*(4), 762–769. https://doi.org/fm9q8z

Mićić, P. (2007). *Phenomenology of future Management in top Management Teams*. Leeds Metropolitan University.

Rauch, W. (1979). The decision delphi. *Technological Forecasting and Social Change, 15*(3), 159–169. https://doi.org/fgw5fv

Roberts, E. B. (1969). Exploratory and normative technological forecasting: A critical appraisal. *Technological forecasting, 1*(2), 113–127. https://doi.org/csw7n2

Saritas, O., & Oner, M. A. (2004). Systemic analysis of UK foresight results: Joint application of integrated management model and roadmapping. *Technological Forecasting and Social Change, 71*(1–2), 27–65. https://doi.org/bxt9mf

Sekaran, U., & Bougie, R. (2013). *Research methods for business: A skill-building approach* (6th ed.). Wiley.

TFAMW Group. (2004). Technology futures analysis: Toward integration of the field and new methods. *Technological Forecasting and Social Change, 71*(3), 287–303. https://doi.org/bf465s

von der Gracht, H. A., & Darkow, I. L. (2010). Scenarios for the logistics services industry: A Delphi-based analysis for 2025. *International Journal of Production Economics, 127*(1), 46–59.

von Hippel, E., Franke, N., & Prügl, R. (2009). Pyramiding: Efficient search for rare subjects. *Research Policy, 38*(9), 1397–1406.

Big Picture of Next Generation Manufacturing

Sebastian Pütz, Marc Van Dyck, Dirk Lüttgens, and Alexander Mertens

Abstract In our real-time Delphi survey, we present 24 projections for Next Generation Manufacturing. An international set of experts from multiple fields, e.g., engineering, information systems, social sciences, and management, evaluated these projections regarding their likelihood and their impact on manufacturing firms by the year 2030. The experts predict that in the coming decade, we will see a significant increase in the use of production data in the form of digital shadows, which will in turn shape both internal and external processes of manufacturing companies. The quantitative results of the Delphi study show that there is significant disagreement among the experts about the likelihood and impact of several of the projections. The most likely projection is the increased importance of environmental sustainability, while the least likely is the emergence of a central platform provider for Next Generation Manufacturing. The most impactful projections are those related to the roles of digital services, data sharing, hybrid intelligence, and environmental sustainability.

[Abstract generated by machine intelligence with GPT-3. No human intelligence applied.]

1 Overview

In our real-time Delphi survey, we present 24 projections for Next Generation Manufacturing. An international set of experts from multiple fields, e.g., engineering, information systems, social sciences, and management, evaluated these projections regarding their likelihood and their impact on manufacturing firms by the year 2030. The final list of projections is provided in Table 1. The projections are

S. Pütz (✉) · A. Mertens
Institute of Industrial Engineering and Ergonomics, RWTH Aachen University, Aachen, Germany
e-mail: s.puetz@iaw.rwth-aachen.de; a.mertens@iaw.rwth-aachen.de

M. Van Dyck · D. Lüttgens
Institute for Technology and Innovation Management, RWTH Aachen University, Aachen, Germany
e-mail: vandyck@time.rwth-aachen.de; luettgens@time.rwth-aachen.de

© The Author(s), under exclusive license to Springer Nature Switzerland AG 2022
F. T. Piller et al. (eds.), *Forecasting Next Generation Manufacturing*, Contributions to Management Science, https://doi.org/10.1007/978-3-031-07734-0_3

clustered according to the framework adopted from Gawer (2014), whereby projections (P1)–(P6) belong to the *governance* dimension, (P7)–(P13) to *organization*, (P14)–(P17) to *capabilities*, (P18)–(P21) to *interfaces*, and (P22)–(P24) to the additional dimension of *resilience*. The projections were initially developed by interdisciplinary workshop groups and were subsequently refined and filtered based on a pre-test with a group of 13 experts (for more details on the methodology, see chapter "Applying the Real-Time Delphi Method to Next Generation Manufacturing").

Out of a total of 35 experts from the industry and academia, 29 completed the survey in full, while 6 completed the survey in part. All projections were assessed by 17 to 23 experts from the industry and by 12 experts from the academia. Each projection received ratings from 21 to 26 experts from Germany and from 8 or 9 experts from the rest of the world. This resulted in a total of 1930 quantitative estimations which were further supported by 629 qualitative arguments. As a result, we give an overview of Next Generation Manufacturing based on the experts' quantitative assessments.

2   Expert Assessments

Overall, the experts' assessments of the projections show that the role of digital shadows in production by 2030 is still the subject of controversial debate (see Table 1). To determine whether the experts reached a consensus in estimating the probability of a projection, we used the interquartile range (*IQR*) of their estimates. To improve comparability with prior reports of Delphi studies in the literature, we downscaled the percentage estimates by dividing them by 10 before calculating the *IQR*s and used an *IQR* of 2 or less as the criterion for consensus (see, e.g., Jiang et al., 2017; Scheibe et al., 1975; von der Gracht, 2012; von der Gracht & Darkow, 2010). Based on this threshold, the experts reached a final consensus on the probability of only 4 of the 24 projections developed. The topics on which the experts agreed are the role of subscription models for production machinery (P1), the reduction of labor through AI-based software and robots (P13), the increasing importance of environmental sustainability of production (P15), and the decentralization of supply chains (P22). The experts concurred that the increased role of environmental sustainability is particularly likely (67%), while for the other three projections, they consistently indicated medium probabilities (between 52% and 58%). The projections that resulted in the highest level of dissent between the experts are the emergence of a central platform provider as the operating system for the Industrial Internet of Things (P4), the mediating role of platforms for data sharing (P5), the implementation of adequate measures for protecting employees' privacy (P12), and plant management from home (P21). For all four of these projections, the probability estimates yielded *IQR*s equal to or greater than 4.

Figure 1 displays the average estimates for probability and firm impact for all 24 projections. Starting by focusing on the probability dimension, the projections

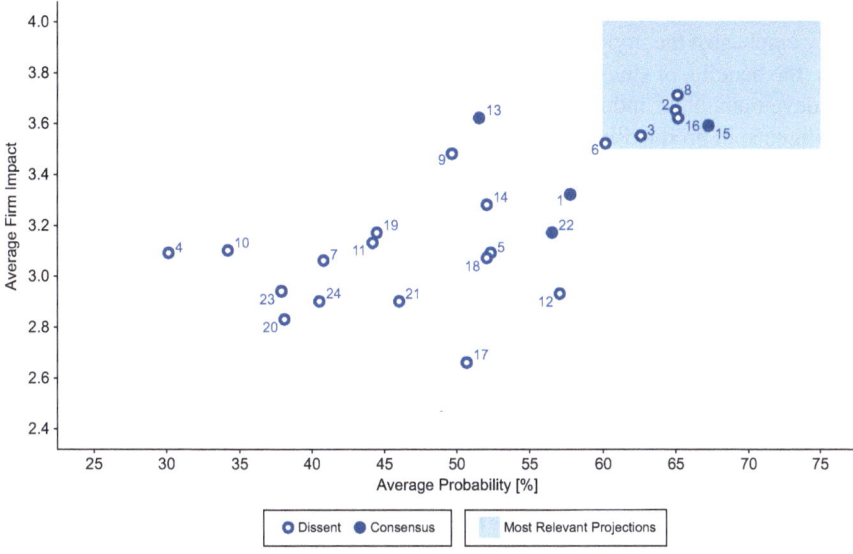

Fig. 1 Expert assessments of 24 projections for Next Generation Manufacturing

show a wide spread of average estimated probabilities, ranging from 30% to 67%, with a mean (M) of 52% and a standard deviation (SD) of 11%. Closely following the projection on environmental sustainability (P15), which was estimated to be the most likely, the next most likely projections are the rising importance of digital services (P2), hybrid intelligence (P8), and full transparency of production systems based on digital twins (P16), all with an estimated probability of 65%. In contrast, the experts considered the emergence of a central platform provider for Next Generation Manufacturing (P4: 30%) and the upheaval of current management structures by virtue of AI-based decision systems (P10: 34%) to be the least likely to occur by 2030. Regarding the potential impact of the projections on firms, the experts' average ratings on the 5-point scale varied between 2.66 and 3.71 ($M = 3.22$, $SD = 0.30$). Consistent with their high probability ratings, projections (P2) and (P8) also received the highest average impact scores (3.65 and 3.7), whereas the development of new multidisciplinary university degree programs (P17: 2.66) and the possibility for production workers to operate their workstations from home (P20: 2.83) are expected to have the lowest impact on manufacturing companies.

Considering the estimates of both probability and firm impact, six projections that the experts consider the most relevant for the future of production emerge. They are highlighted by the colored area in Fig. 1. The rationale for using the selected post hoc cut-off values of probability estimates above 60% and firm impact ratings above 3.5 is that the resulting 6 projections are assessed as being more likely than any of the other 18 projections and represent 6 of the 7 most impactful projections. The only projection that has a higher estimated firm impact than some of the six selected projections is rated as considerably less likely. Therefore, the results of this Delphi study emphasize the importance of considering the increasing role of digital services (P2), the benefits of sharing usage and production data with business partners (P3), the development of industrial data protection regulations (P6), the role of hybrid intelligence in production decision-making (P8), the environmental sustainability of production (P15), and the full transparency of production systems based on digital twins (P16) in the strategic planning of production firms and in related future research.

3   Comparison Between Subgroups of Experts

Although the further analysis and discussion of the expert assessments in this book focus on the full expert sample, we present here some additional insights into the experts' perspectives by comparing the assessments of different types of experts. Figure 2a illustrates the differences in assessments between academic and industry experts. When calculating the *IQRs* of the probability estimates for consensus identification within the subgroups, both groups show a consensus for projection (P15), consistent with the findings for the full sample. However, both groups also yielded a consensus for projection (P4), which yielded one of the highest levels of dissent when looking at the full sample. The observed dissent can thus be attributed to the different perspectives of the two groups of experts, which are internally consistent, with academics estimating a higher probability than industry representatives. The academics also yielded a consensus on the probability of introducing collaborative robots in production (P7).

While the mean average probability ratings of the academic ($M = 53\%$, $SD = 10\%$) and industry ($M = 50\%$, $SD = 11\%$) experts are similar, experts from the academia rated 17 of the 24 projections as more likely than their industry counterparts. That said, a Wilcoxon signed-rank test did not show a significant difference in the probability ratings between the two groups ($p = 0.056$, $r = 0.275$). Individual projections that are considered more likely by the academics include the introduction of collaborative robots (P7), the disruptive effect of AI-based decision systems on established leadership structures (P10), and the decentralization of supply chains (P22). In contrast, the industry experts attributed a higher probability to the implementation of digital shadows of production workers (P11) and the introduction of adequate anonymization procedures for the protection of employees' personal rights (P12) than the academic experts. Regarding the firm impact ratings, academics ($M = 3.30$, $SD = 0.33$) and industry members ($M = 3.19$, $SD = 0.34$) also yielded similar mean average rating, with the academics rating the firm impact of the projection higher than industry members in 15 of 24 cases. The difference between the ratings of the two groups was again not significant

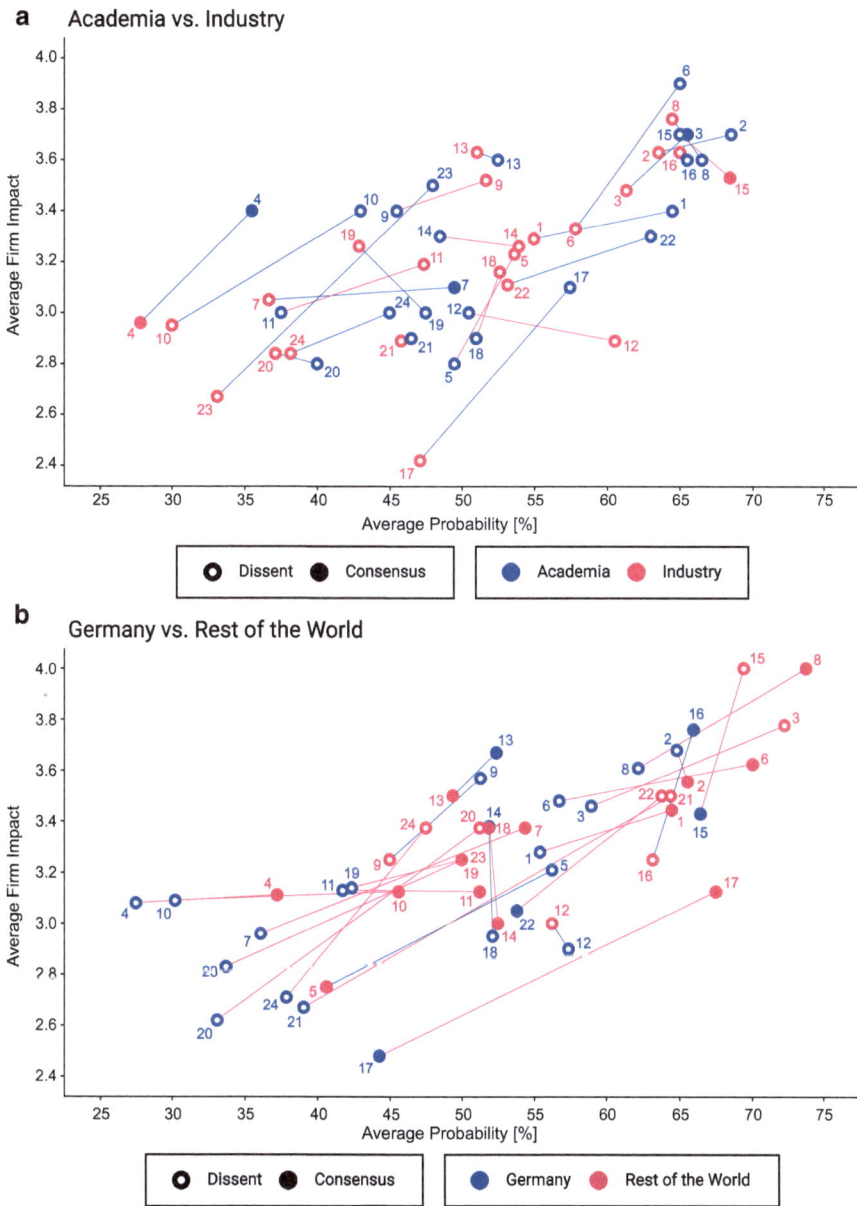

Fig. 2 Expert assessments of 24 projections for Next Generation Manufacturing differentiated between experts from the academia and from the industry (**a**) and between experts from Germany and the rest of the world (**b**). The connecting line between the two assessments for each projection is colored according to the group with the higher probability rating

($p = 0.145$, $r = 0.210$). The largest differences between the two groups, corresponding to higher impact assessments by the academic experts, are for the projections of new multidisciplinary university degree programs (P17) and increasing production costs due to more regional production and higher inventory levels (P23). Conversely, the industry experts considered the introduction of platforms as mediators in data sharing (P5) in particular as more impactful than the academic experts.

Switching to the comparison between experts from Germany and from the rest of the world (see Fig. 2b), this division of the full expert sample highlights the higher levels of consensus among the experts within the respective subsamples. Whereas the German experts yielded a consensus for projections (P13), (P15), (P16), (P17), and (P22), the experts from other countries did so for projections (P1), (P2), (P4), (P5), (P6), (P7), (P8), (P10), (P11), (P13), (P14), (P17), (P18), and (P23). This observation may indicate the importance of experts' regional background and cultural experiences for their predictions on the future of production. However, the small sample sizes of the subsamples should be considered, as they affect both the informative value of the *IQR* as a measure of dispersion and the overall generalizability of the inferred conclusion.

In terms of probability estimates, the experts from Germany ($M = 49\%$, $SD = 12\%$) showed a lower mean average estimation than the other experts ($M = 57\%$, $SD = 10\%$), providing lower estimates for 18 of the 24 projections. Based on the performed Wilcoxon signed-rank test, this difference between the average probability estimates of the two groups reaches statistical significance ($p < 0.01$, $r = 0.456$). Future developments that the German experts consider particularly less likely are the introduction of new multidisciplinary university degree programs (P17) and plant management from home (P21). In contrast, the emergence of platform providers as mediators for data sharing (P5) is the projection with the largest difference in average probability, with the German experts estimating its probability to be higher. In addition, the two groups also differ significantly in their average estimates of the impact of the projections on firms ($p < 0.05$, $r = 0.306$). For 17 of the 24 projections, the German experts ($M = 3.17$, $SD = 0.37$) assessed the firm impact as being lower than the experts from the rest of the world did ($M = 3.36$, $SD = 0.30$). Exemplary projections that yielded high differences in impact estimates between the two groups are (P5), (P16), (P20), and (P21). Whereas the German experts estimated lower firm impacts for production workers (P20) and plant managers (P21) working from home, they assessed platform providers as mediators for data sharing (P5) and full transparency of production systems based on digital twins (P16) as being more impactful than the other experts.

4 Summary

The experts project that in the coming decade, we will see a significant increase in the use of production data in the form of digital shadows, which will in turn shape both internal and external processes of manufacturing companies. The experts ascribe a high probability and a high firm impact to the visions of achieving full transparency of production processes via digital twins of production machines, production lines, and plant engineering and operation. This progress in creating comprehensive datasets comprising information on all relevant aspects of production will create vast opportunities, from improving decision-making through AI-based assistance to creating new business models by sharing data between companies and providing newly developed digital services. However, although the experts provided positive assessments for the central vision of Next Generation Manufacturing, their responses also emphasize that there is still considerable uncertainty about how exactly the deployment of digital shadows will impact the production landscape, as shown by the dissent among them for most projections. These differences in the experts' assessments can be partially attributed to their different professional and regional backgrounds. This observation highlights both the importance of using a diverse panel of experts to forecast Next Generation Manufacturing and the need for further research on the differences between the perspectives of various groups of experts. Indeed, the latter is especially important, as the opinions and expectations of relevant stakeholders will have a direct influence on future developments, with significant differences between stakeholder groups potentially leading to tensions or divergent developments in different geographic and economic areas. To conclude, data-based optimization and value creation will be a central part of Next Generation Manufacturing, though the details are still difficult to predict.

Acknowledgment Funded by the Deutsche Forschungsgemeinschaft (DFG, German Research Foundation) under Germany's Excellence Strategy – EXC-2023 Internet of Production – 390621612.

Appendix

Table 1 List of the 24 projections with descriptive statistics of the expert assessments

	P#	Projection	Description	N	Probability of occurrence (in %)			Firm impact
					IQR	M	SD	M
Governance	1	Subscription models	In 2030, subscription models for production machines will be the new industry standard, fulfilling an assured performance level based on real-time usage data in return for a periodic payment	35	2.0	57.79	19.22	3.32
	2	Digital services	In 2030, for production machinery and other hardware assets, e.g., tractors, equipment, etc., competition will shift from hardware capabilities and functionality to differentiation by (digital) services, supplementing the traditional transactional business logic with a data-driven business model	35	3.0	65.00	19.85	3.65
	3	Data sharing	In 2030, organizations that share usage and production data with suppliers, customers, and other partners will obtain a competitive advantage over organizations that do not share this data	33	3.0	62.58	21.89	3.55
	4	Central platform	In 2030, one central platform provider will serve as the operating system for the Industrial Internet of Things, enabling them to make use of data by integrating machine manufacturers and complementary service providers and capturing the greatest share of the value created	33	4.0	30.15	18.93	3.09
	5	Data mediator	In 2030, platform orchestrators or dedicated third-party providers will mediate data sharing between all actors involved in a production network	32	4.1	52.34	24.65	3.09
	6	Industrial GDPR	In 2030, industrial data protection regulations (like a special GDPR – General Data Protection Regulation for Business-to-Business) will govern the application of data-based digital services	31	2.5	60.16	22.52	3.52

Organization	7	Autonomous robots	In 2030, collaborative robots that move autonomously on the shop floor and interact directly with humans will have replaced most conventional robots that only interact in protected cells	31	3.5	40.81	25.72	3.06
	8	Hybrid intelligence	In 2030, strategic production decisions will be executed with close interaction between humans and AI-based algorithms ("hybrid intelligence")	30	3.0	65.13	23.66	3.71
	9	AI-based assistants	In 2030, operative production decisions will no longer lie with people, as they will be made by AI-based decision-making agents	30	3.5	49.68	22.61	3.48
	10	New leadership	In 2030, AI-based decision systems will have changed our current understanding of management completely, increasingly eliminating hierarchies and leadership based on human interactions	30	3.0	34.19	20.25	3.10
	11	Human digital twins	In 2030, a full digital twin of each production worker and all of her/his operations will be available and will become a valuable tool for production planning and optimization by reflecting their workload, their stress, and also their need for training in real time	30	3.8	44.19	23.80	3.13
	12	Employees' rights	In 2030, adequate anonymization procedures for the protection of employees' personal rights will have been introduced for firms that collect data on personal performance and work patterns in the form of digital twins of their employees	29	5.0	57.07	26.01	2.93
	13	Workforce reduction	In 2030, AI-based software and robots will have reduced a firm's workforce significantly	29	2.0	51.55	18.85	3.62
Capabilities	14	Expert knowledge	In 2030, implicit expert knowledge which traditionally could only be gained through experience will increasingly be explicitly preserved in the form of digital models, interactive guides, or instructions and facilitated by technologies like augmented or virtual reality. As a result, this knowledge will also be made available to novices and will eliminate the dependency on experienced production employees	29	3.5	52.07	20.95	3.28
	15	Environmental sustainability	In 2030, environmental sustainability of production will have increased significantly compared to today	29	2.0	67.24	19.14	3.59
	16	Production transparency	In 2030, full transparency based on a complete digital twin of all production machines, lines, and plant engineering and a complete digital twin of their operations will increase production efficiency significantly	29	3.0	65.17	20.82	3.62

(continued)

Table 1 (continued)

	P#	Projection	Description	N	Probability of occurrence (in %)			Firm impact
					IQR	M	SD	M
	17	University degrees	In 2030, the application of biological principles (e.g., cybernetics, biomimicry) of manufacturing will have created a demand for new multidisciplinary university degrees covering engineering, the life sciences, and computer science	29	3.0	50.69	23.48	2.66
Interfaces	18	Implicit interfaces	In 2030, human-machine interaction will have evolved away from explicit interaction, where the human operator has full control of the actions of the production system's entities, toward implicit interaction, where the system automatically adapts to the human operator's behavior by detecting and predicting their actions and modifying these actions accordingly	29	3.0	52.07	17.93	3.07
	19	Open interfaces	In 2030, regulatory requirements will demand open and standardized interfaces for data exchange for all kinds of manufacturing equipment	29	3.6	44.48	21.77	3.17
	20	Production from home	In 2030, production employees will operate their workstation from their home, controlling, for example, remotely operated robots	29	3.0	38.10	24.93	2.83
	21	Plant management from home	In 2030, plant directors will be able to manage multiple factories centrally from their home due to the complete and real-time transparency of all the operations in a digital system	29	4.0	46.03	25.61	2.90
Resilience	22	Decentralization	In 2030, supply chains will have become more decentralized, with production and sourcing moving closer to the end customer to cope better with global crises (e.g., pandemics)	29	2.0	56.55	21.34	3.17
	23	Production costs	In 2030, production costs will have increased substantially due to more regional production and higher inventory levels to cope with global crises (e.g., pandemics)	30	2.5	37.90	20.37	2.94
	24	Production resilience	In 2030, AI-based decision systems will enable greater resilience of production networks in the event of a global crisis (e.g., a pandemic)	29	2.9	40.52	19.88	2.90

References

Gawer, A. (2014). Bridging differing perspectives on technological platforms: Toward an integrative framework. *Research Policy, 43*(7), 1239–1249. https://doi.org/gc8sc5

Jiang, R., Kleer, R., & Piller, F. T. (2017). Predicting the future of additive manufacturing: A Delphi study on economic and societal implications of 3D printing for 2030. *Technological Forecasting and Social Change, 117*, 84–97. https://doi.org/ghzgpp

Scheibe, M., Skutsch, M., & Schofer, J. (1975). Experiments in Delphi methodology. In H. A. Linstone & M. Turoff (Eds.), *The Delphi method — Techniques and applications* (pp. 262–287). Addison-Wesley.

von der Gracht, H. A. (2012). Consensus measurement in Delphi studies: Review and implications for future quality assurance. *Technological Forecasting and Social Change, 79*(8), 1525–1536. https://doi.org/gddk63

von der Gracht, H. A., & Darkow, I. L. (2010). Scenarios for the logistics services industry: A Delphi-based analysis for 2025. *International Journal of Production Economics, 127*(1), 46–59. https://doi.org/b2dkb6

Governance Structures in Next Generation Manufacturing

Christian Brecher, Matthias Jarke, Frank T. Piller, Günther Schuh,
Annika Becker, Florian Brillowski, Ester Christou, István Koren,
Maximilian Kuhn, Dirk Lüttgens, Marc Van Dyck, and Marian Wiesch

Abstract Next Generation Manufacturing describes a vision of an open network of sensors, assets, products, and actors which are not restricted to a focal organization or a closed supply chain. A core principle of the digital shadow is that it collects and shares data about the usage of products within and across organizations, allowing them to optimize operations, investment decisions, innovation processes, or the generation of new products. Sharing of usage data, however, requires new forms of governance, both internally and externally. Given the high uncertainty in the likelihood of occurrence and the technical, economic, and societal impacts of these

C. Brecher · G. Schuh · A. Becker · M. Kuhn · M. Wiesch
Laboratory for Machine Tools and Production Engineering (WZL), RWTH Aachen University, Aachen, Germany
e-mail: c.brecher@wzl.rwth-aachen.de; g.schuh@wzl.rwth-aachen.de; a.becker@wzl.rwth-aachen.de; m.kuhn@wzl.rwth-aachen.de; m.wiesch@wzl.rwth-aachen.de

M. Jarke
Chair of Computer Science 5 – Information Systems and Databases, RWTH Aachen University, Aachen, Germany
e-mail: jarke@informatik.rwth-aachen.de

F. T. Piller
Institute for Technology and Innovation Management, RWTH Aachen University, Aachen, Germany

Institute for Business Cybernetics (IfU) e.V. at RWTH Aachen, Aachen, Germany
e-mail: piller@time.rwth-aachen.de

F. Brillowski
Institut für Textiltechnik, RWTH Aachen University, Aachen, Germany
e-mail: florian.brillowski@ita.rwth-aachen.de

E. Christou · D. Lüttgens · M. Van Dyck (✉)
Institute for Technology and Innovation Management, RWTH Aachen University, Aachen, Germany
e-mail: christou@time.rwth-aachen.de; luettgens@time.rwth-aachen.de; vandyck@time.rwth-aachen.de

I. Koren
Chair of Process and Data Science, RWTH Aachen University, Aachen, Germany
e-mail: koren@pads.rwth-aachen.de

F. T. Piller et al. (eds.), *Forecasting Next Generation Manufacturing*, Contributions to Management Science, https://doi.org/10.1007/978-3-031-07734-0_4

concepts, we conducted a technology foresight study, in the form of a real-time Delphi analysis, to derive reliable future scenarios featuring the next generation of manufacturing systems. This chapter presents the governance dimension and describes each projection in detail, offering current case study examples and discussing related research, as well as implications for policy makers and firms. For example, according to the experts, subscription models for production machines will be the new industry standard by 2030. This is due to the changing needs of manufacturers and customers, as well as the impacts of digitization and Industry 4.0. Customers would benefit from guaranteed machine availability and lower investment costs, while manufacturers would benefit from increased customer satisfaction and longer-term business relationships.

[Abstract generated by machine intelligence with GPT-3. No human intelligence applied.]

1   Introduction

Next Generation Manufacturing describes a vision of an open network of sensors, assets, products, and actors which are not restricted to a focal organization or a closed supply chain. A core principle of the digital shadow is that it collects and shares data about the usage of products within and across organizations, allowing them to optimize operations, investment decisions, innovation processes, or the generation of new products. Sharing of usage data, however, requires new forms of governance, both internally and externally. Governance refers here to the reconciliation of the interests of all internal and external stakeholders (Gawer, 2014). Internally, governance systems need to reflect changes in organizational structures and processes. They need to create acceptance by all stakeholders and ensure security, privacy, and ethical behavior. Externally, governance needs to balance the need for openness in order to integrate third parties with each actor's desire for control. The internal and external governance systems also need to reconcile contradicting requirements. For instance, they need to ensure internal privacy while enabling a certain degree of external openness.

To create value based on a digital shadow, data must be exchanged with third-party complementors – independent but interdependent actors (Jacobides et al., 2018) – who innovate based on the data, e.g., in the form of dedicated "apps" (Gawer, 2014; Parker & Van Alstyne, 2018). These exchanges resemble the dynamics in platform markets (e.g., Parker et al., 2016). Collaborative innovation needs to be coordinated and federated, which can be done through different governance modes (Gawer, 2014). While basic mechanisms of value creation in platform markets are understood, dedicated research in the context of industrial data applications is lacking, as is on work on value capture.

A central challenge for governance is finding the right degree of openness to enable both value creation and capture (Boudreau, 2010; West, 2003). The more open data is exchanged, the more innovative input can be provided by third parties. However, the less control the digital shadow provider has, the less value the focal organization can capture. In addition, governance can be implemented differently. It can be either centrally organized by the provider of the digital shadow or decentrally distributed between multiple organizations (Eisenmann et al., 2006). These governance challenges are exacerbated by two factors: first, ecosystems built on digital shadows are newly emerging, and it is not yet clear what future complements and third-party actors are required (Dattée et al., 2018); and second, digital shadows enable completely new value propositions, such as a shift from acquiring reliable industrial equipment to paying for access to outcome-based services, with performance assessed through the data generated by the equipment (Iansiti & Lakhani, 2020; Porter & Heppelmann, 2015).

Using a novel real-time Delphi approach (see chapter "Applying the Real-Time Delphi Method to Next Generation Manufacturing" for a presentation of the method and the sample and chapter "Big Picture of Next Generation Manufacturing" for an overview of the results), we developed propositions for different scenarios within Next Generation Manufacturing in 2030. As suggested by Gawer (2014), we used an integrative framework for platforms, distinguishing four dimensions: governance (e.g., open forms of collaboration, this chapter), organization (e.g., boundaries and decision-making; see chapter "Organization Routines in Next Generation Manufacturing"), capabilities (e.g., hybrid intelligence; see chapter "Capability Configuration in Next Generation Manufacturing"), and interfaces (e.g., open APIs and human-machine interfaces; see chapter "Interface Design in Next Generation Manufacturing"). In addition, and influenced by our shared experiences during the COVID-19 pandemic, we added a fifth cluster of propositions addressing the need for resilience in future digital manufacturing systems (see chapter "Resilience Drivers in Next Generation Manufacturing"). We provide a set of 24 validated projections based on 1930 quantitative estimations and 629 qualitative arguments from 35 industrial and academic experts from Europe, North America, and Asia. In so doing, we deliver a basis on which to substantiate academic discussions and which can support firm decision-making on future technological developments and economic implications that go beyond current speculations and siloed research.

To cover all perspectives on governance challenges, we developed six projections. Projections 1 (subscription models) and 2 (digital services) address potential changes in value propositions. We explored whether new types of ownership models and closer exchanges with users are likely and which resources will provide competitive advantages in the future. Projections 3 (data sharing), 4 (central platform), and 5 (data mediator) explore different implementation modes of governance – from central to decentral – and the incentives required for firms to share data. Lastly, Projection 6 (industrial GDPR) addresses relevant security and privacy concerns within and across organizations (see Fig. 1).

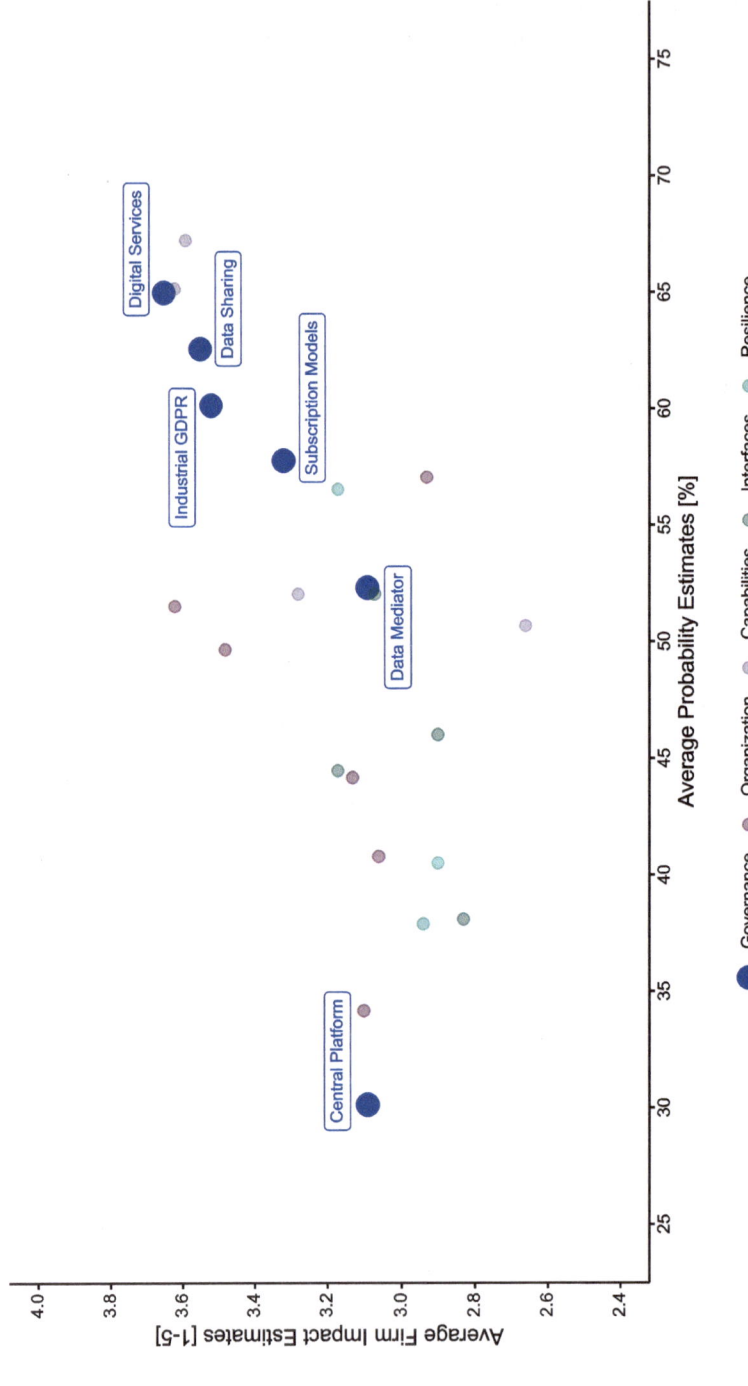

Fig. 1 Expert assessment on governance structures in Next Generation Manufacturing (see chapter "Big Picture of Next Generation Manufacturing" for the full results of the Delphi survey)

2 Projection 1: Subscription Models

The first projection in the governance dimension says that in 2030, subscription models for production machines will be the new industry standard, fulfilling an assured performance level based on real-time usage data in return for a periodic payment. The results of the study show a high probability of occurrence of this projection, with a mean probability of 57.79%. The low interquartile range (*IQR*), with a value of 2.00, shows that a consensus could be found among the experts about this probability of occurrence. At the same time, the experts rated the impact of this projection on the industry and its companies as relatively high, with a mean value of 3.32.

The high probability of occurrence and the high impact on companies can be explained by changing needs on the part of manufacturers and customers, as well as by overarching trends and technical developments in the machinery and plant engineering sector. First, the intensity of competition in machinery and plant engineering is increasing due to globalization (Schuh et al., 2017). Especially in high-wage countries, differentiation based on cost-effective products is no longer expedient for companies in this sector (Schuh et al., 2020). In this context, the combination of products and services in so-called product-service systems (PSS) appears to be a promising approach. PSS offer the possibility to differentiate oneself from competitors, generate comparatively high margins, and increase customer loyalty. At the same time, digitization and Industry 4.0 are finding their way into the field of machinery and plant engineering, enabling the development of new, innovative business models (Meier & Uhlmann, 2012).

In this context, a promising type of business model is subscription models (Schuh et al., 2019). A subscription model is defined by the continuous delivery of a value proposition in return for a recurring periodic fee (Gassmann et al., 2013; Rappa, 2004; McCarthy et al., 2017). In contrast to one-time, product-oriented transactions in classic transactional business models, providers of subscription models can use them to meet their customers' needs continuously and to build long-term business relationships (Tzuo, 2018). Subscription models with these characteristics have been implemented in different industries going back several centuries.

In the early sixteenth century, European map publishers used subscription models to give continuous updates on conquered empires (Rudolph et al., 2017). Since 1600, subscription models been used to provide access to newspapers and books (Warrillow, 2015). In the B2B sector, the aircraft turbine manufacturer Rolls-Royce established one of the first and best-known subscription models in the early 1960s. Their "Power-by-the-Hour" business model focuses on the usage of their turbines instead of their purchase as a product (Rolls Royce, 2012). Driven by continuous digitization, subscription models have been established in the IT (e.g., Salesforce) and multimedia (e.g., Netflix) sectors. Based on usage data from their subscription model, Netflix gained a much better understanding of their customer's needs and transformed itself from a DVD distributor into one of the most successful producers of television series and movies (Schuh et al., 2019). With Heidelberg Subscription, the German printing machine manufacturer Heidelberger Druckmaschinen was one of the first companies in the machinery and plant engineering sector to offer a business model in which customers pay for a defined monthly print volume instead of paying for the printing machine itself (Riesener et al., 2020).

In the machinery and plant engineering sector, the potential of subscription models is broad and can help secure Germany as a high-wage location in the long term. There are numerous advantages for both customers and manufacturers. For example, customers benefit from the manufacturers' agreed use-oriented performance promise. This ensures, for example, guaranteed machine availability and a certain productivity of the machine. Detailed knowledge of the machines based on usage data also enables manufacturers to deliver a certain value proposition. Through the close connection between manufacturer and customer during the usage phase, the potential improvements to the machine can be jointly identified, and productivity continuously increased. In addition, customers can benefit from greater financial flexibility due to lower investment costs. Machines can be purchased for a purchase price and an additional fixed or usage-based fee (Schuh et al., 2020).

There are also many advantages for manufacturers. First, it is scientifically proven that customers show an increased willingness to pay due to the additional benefits and the integrated solution to their problems (Tukker, 2004). Greater customer satisfaction also leads to long-term business relationships and secures payment flows in the long term, which can lead to significantly higher profitability compared to traditional sales (Schuh et al., 2020).

The high impact on machinery and plant engineering companies identified in the study can be explained by the actions these companies need to take to implement subscription models. Riesener et al. (2020) suggest an iterative approach for this. Accordingly, the implementation process starts with the strategic conception of the subscription model. Subsequently, concrete insights about customers must be gained and used to specify the subscription model. Based on this, a PSS is set up and usage data are analyzed, and the scope of service and the remaining framework conditions of the subscription model are adjusted if necessary (Riesener et al., 2020).

An example of a subscription business model for production machines is provided by the company DMG Mori AG, who are introducing a new digital business model in the form of a subscription model with their PAYZR – Pay with Zero Risk – product. Customers can subscribe to machines instead of purchasing them. Consequently, there is no investment risk or down payments, but full financial flexibility, cost and price transparency, and, as a result, a high level of planning security. PAYZR is offered as Equipment-as-a-Service or as Software-as-a-Service. With Equipment-as-a-Service, customers pay a basic monthly fee and a usage-based fee per spindle hour. PAYZR can be easily accessed via several digital channels, such as the website or the "my DMG Mori" customer portal (DMG Mori, 2021).

3   Projection 2: Digital Services

Besides subscription models, additional opportunities are being seriously explored by the industry, indicating that a transformation of business practices is imminent. The expert survey revealed that competition for production machinery, e.g., machine tools, and other hardware assets, e.g., tractors or equipment, will with a high probability ($M = 65.00\%$) and a high firm impact ($M = 3.65$) shift from differentiation through hardware capabilities and functionality to differentiation through digital services. Nevertheless, the expert evaluations in the survey show disagreements between the experts' assessments, especially surrounding the future importance of hardware and mechanical functionality associated with technical system complexity in competition with the position of software as the basis for digital services ($IQR = 3.00$). However, the experts agree that it will be important to remain competitive in terms of hardware while aiming for a good mix of both physical and digital goods (e.g., licensing, digital models, and intelligent properties). Even if a trend toward expanding digital services in the context of comprehensive business model innovation is discernible in some industrial sectors, the discussion on the use of digital services among the experts was characterized by a bottom-up approach, with a focus on technology and standards. The research and implementation of systematic business model innovation processes in the context of Industry 4.0 will drive this development.

Due to technological leaps and societal changes, a significant increase in digital goods is currently becoming apparent. This trend started with massively multiplayer online role-playing games (MMORPGs) and sports games, in which cosmetic or temporary items of a digital nature can be purchased with real money. Nowadays, sports game publishers make most of their revenue from digital goods that lose their value within a single year (Brillowski et al., 2021). Evidence of this can also be found in production, of which the most prominent example is additive manufacturing, where the value of digital models exceeds the value of products and production machinery (Chekurov et al., 2018; Korbel et al., 2019). In other production areas, profits are currently being made from the licensing of software or maintenance contracts.

Machine tool manufacturers have also recognized the potential of digital services: K.K. Makino Furaisu Seisakujo offers its customers a holistic monitoring and control tool, the so-called MHmax (Makino Health Maximizer), which continuously monitors the condition of a machine tool using integrated sensors. Makino offers its customers three connectivity levels. While level 1 is limited to individual machines, level 2 allows cross-company analysis and access to all data within the company network. Level 3 includes connection to the Makino cloud. This gives Makino access to machine data and enables it to achieve improved analysis based on extensive historical data. The state of the subsystems can be viewed continuously via dashboards. In addition, warnings and malfunctions alerts, as well as essential information for early, proactive intervention, are provided. The overarching analysis of historical data combined with the manufacturer's expertise enables significantly more accurate predictions which can be used as the basis for fast and predictive intervention decisions (Brecher et al., 2021).

The example of MHmax shows the current efforts in the machine tool industry to offer digital services in the form of tailored proposals, depending on the customer's use of the individual hardware and on the network of production units. For example, these proposals may allow the optimal and resource-efficient use of the tools, as well as enabling productivity-increasing measures for individual processes through the availability and evaluation of process data. These efforts highlight the disagreement described among the individual experts about whether hardware capabilities and functionality will transform fully into (digital) services. Breaking the projection down into a combination of hardware functionality and software as the basis for (digital) services, the experts in the survey see a combination of both being important and revealing undiscovered potential in relation to improved ergonomics, efficiency, and flexibility in the future. The importance of the topic is reflected in the firm impact as well as in the probability of occurrence; the disagreement of some of the experts comes from the hardness of the statement in its projection of a complete shift. Until recently, it was not necessary to address production data-related issues to remain competitive, but technological leaps and societal and technological changes mean an emphasis on digital goods or services will be required to stay competitive in 2030. In this area, legislative, software-, and hardware-related questions and research opportunities arise in regard to customs clearance of digital goods, licensing possibilities, and open access. In addition to the endless possibilities, it will also be necessary to address problems related to data leaks and cyber piracy in order to safeguard knowledge and prevent production sites in high-wage countries from being jeopardized. However, to establish digital services long term within a producing company's portfolio, novel business models that emphasize data acquisition and sharing should be the focus of research that follows a systematic business model innovation process in the context of Industry 4.0 (Burmeister et al., 2016).

4   Projection 3: Data Sharing

The survey indicates that it is highly likely that organizations that share usage and production data with suppliers, customers, and other partners will obtain a competitive advantage over organizations that do not share this data ($M = 62.58\%$). The experts predict that the projection will have a high impact if it materializes ($M = 3.55$). Firms' core knowledge will change decisively, and new features will be available due to data sharing. The experts particularly highlight the benefits of monitoring and improving supply chains and consider data sharing as a prerequisite for innovative solutions such as predictive maintenance. Nevertheless, the experts did not reach a consensus, as the value of data sharing was considered questionable by some ($IQR = 3.00$).

In an organizational context, data can be considered as a production resource, just like human, machinery, capital, and other such resources (Barney, 1991; Levitin &

Redman, 1998). In recent decades, most data gathered by organizations did not leave the internal system (Fitzgerald, 2013). Conservative firms might be afraid of sharing data as their data are a valuable resource that other companies could potentially use and gain benefits from. For managers and employees, increased data sharing requires new capabilities to cope with data-based business models and new collaboration modes with external organizations.

Yet in recent years, interconnected businesses have become more and more prevalent. As an example, platform-based business models with data sharing as a core aspect of how they capture value already exist, especially in the online gaming industry (Boudreau & Jeppesen, 2015) and social networks (Li & Agarwal, 2017).

For firms in the industrial sector, sharing data also holds potential for capturing competitive advantages. Data sharing enables competitive supply chains, better usage of machines, and the transformation to digital platform business models. The competitiveness of supply chains no longer relies on how companies design their contracts with regard to making data available for other players in multisided markets. Rather, it relies on how companies agree to share data within already existing business models (Huttunen et al., 2019).

Data sharing plays a key role in efficient supply chains because it matches supply and demand to reduce costs by increasing resource utilization, thereby reducing excess inventory and loss of profits from stockouts (Stefansson, 2002). Researchers suggest that the application of information technologies sharing data has substantially lowered the time and cost required to process an order, resulting in improvements in supply chain performance (Clark & Hammond, 1997).

Data sharing can also enable better usage of machines: Condition Based Services (CBS) by TRUMPF increases the availability and productivity of networked laser systems. CBS evaluates important sensor data and detects risks and potential for improvement. All data (more than 250 sensors per machine) are analyzed centrally by algorithms and TRUMPF experts. This helps to avoid high production downtime costs (typically several 1000 euro per hour). In addition to automatic, proactive warning messages, the system also provides recommendations for optimized operation. Beyond this, the system provides full transparency by aggregating all data in customer-specific dashboards.

The collaboration between DuPont and John Deere serves as an example of exploiting new business models based on data collection and sharing. DuPont and John Deere developed a new business model based on sharing their data. The goal of the collaboration between these two firms is to support decision-making related to planting, harvesting, and field management in order to maximize crop yields. DuPont's key activity is agricultural consulting and selling seeds. John Deere is one of the leading manufacturers of agricultural machinery, with a value proposition of farming equipment outfitted with sensors, GPS, and wireless transmission technology. Both firms have farmers as their main customers. John Deere's farming equipment gathers data on crop yields, moisture, and location and is sent to DuPont, who use this data in their farming software that supports decision-making. DuPont actively integrates John Deere's Data into its value proposition. Products from both companies gain value by sharing data, and better services can be provided to customers, creating a competitive advantage (Deloitte, 2014).

5 Projection 4: Central Platform

The survey indicates high uncertainty as to whether one central platform provider will serve as the operating system for Next Generation Manufacturing, connecting machines, complementary assets, data, and service providers from different organizations ($IQR = 4.00$). The experts deemed this projection rather unlikely due to the heterogeneity of both competition and machines and existing legacy technology ($M = 30.15\%$). At the same time, the survey showed a medium to high impact if the projection materializes, indicating its potentially transformative character ($M = 3.09$).

Platforms already dominate consumer markets, as illustrated by firms like Alibaba, Uber, and Google, who have shifted from selling products and services toward facilitating economic exchanges between two or more (related) user groups, realizing network effects (Zhao et al., 2020). In these platform markets, a focal organization usually orchestrates all actors through a central platform, and over time, winner-takes-all (WTA) scenarios evolve when the market shifts toward a dominant platform due to strong network effects (Eisenmann et al., 2006; Cennamo & Santalo, 2013; Jacobides et al., 2018). The provider of the central platform usually captures the most significant share of the value (Zhu & Iansiti, 2019).

Platforms have only recently started to evolve in industrial (BtoB) markets (e.g., Porter & Heppelmann, 2015; Kopalle et al., 2020). Like in consumer markets, many manufacturing firms are trying to build their own platforms to connect machinery providers with service providers, thus enabling complementary innovation based on data and allowing them to capture unprecedented amounts of value (Kopalle et al., 2020). Prominent examples include Siemens' MindSphere, General Electric's Predix, and TRUMPF's AXOOM. In contrast to consumer markets, WTA dynamics do not necessarily lead to one central platform provider. Instead, multiple platforms can co-exist (Piller et al., 2021).

For society, a central platform provider would pose challenges concerning antitrust regulations and monopolistic market structures, since network effects based on data could change the market dynamics (Gregory et al., 2021). Other projections (P3, P5, P6, P19) propound viable solutions. For firms, building a platform on their own is risky (e.g., Yoffie et al., 2019). Following an alliance-driven approach, where the platform (or multiple platforms) is operated under shared leadership, could provide an effective alternative. Initiatives such as the Industrial Data Space (IDS; Otto & Jarke, 2019) and GAIA-X (Braud et al., 2021) suggest potential blueprints for alternative decentralized structures.

6   Projection 5: Data Mediator

The survey shows that there is dissent among respondents on whether data sharing between all actors of a production network will be managed individually or by an external entity (i.e., neutral mediator) ($IQR = 4.10$). It is estimated that there is a medium likelihood ($M = 52.34\%$) that data exchange will be mediated by a platform orchestrator or dedicated third-party services and that if this happens, it will have a medium impact ($M = 3.09$).

In order to submit data analytic queries across organizational boundaries, a technical mapping between the export view (the perspective the data provider is willing to reveal) and the import perspective must be defined. The latter allows the data consumer to embed the query results within its own data infrastructure to process them. Furthermore, such export views need to respect privacy regulations concerning personal data. Traditionally, such mappings are defined as programmed workflows. However, more recently, also logic-based semi-automatic mappings have been proposed. On the export side, popular means of protecting the data provider's confidential knowledge are aggregation, encryption, and fuzzification (Schuh et al., 2022).

In the information systems literature, architectures for data mediation have been proposed as an infrastructure for integrating data from different sources into a central schema since the late 1980s (Jarke et al., 1987), e.g., for data warehouses (Jarke et al., 2002) or multi-party negotiations (Jarke et al., 1987). In Wiederhold's (1992) wrapper-mediator architecture, wrappers are responsible for defining and executing the transformation from source data models to the central model. In contrast, mediators resolve conflicts among contradictory data such as missing information, different times of data capture, and other data quality issues. Especially in the mediation of production data, data provenance is considered particularly relevant (Gleim et al., 2020; Becker et al., 2021).

Considering the important aspect of physical data transport, commercial data warehouses nowadays follow a three-stage ETL (Extract-Transform-Load) architecture. For data exchange, this process must be performed twice, once on the provider side and then in reverse order on the consumer side of a data process. This is mediated by a metadata management tool called a broker. On a much larger scale, commercial brokering platforms like SAP Convergent Mediation collect and analyze highly heterogeneous usage data from many different sources and act as a mediator, distributing the data for usage to diverse organizations. Similar offerings exist from vendors such as IBM Data Integration or AWS Data Exchange. The business advantage of such mediation platforms is that the number of linkages that need to be programmed grows only linearly with the number of sources, rather than quadratically, as was the case when mappings between all pairs of sources were necessary.

Recent research on ontology-based data exchange (Lenzerini, 2019) demonstrates that direct peer-to-peer data exchange without such central platforms can also be a feasible option. It could, in fact, be more efficient than the double work of first transforming source data to a central schema and then transforming that transformed data back to the target schemas on the consumer side. In peer-to-peer data exchange, mappings relate the semantic meaning of data source models and consumer data models to a shared formal knowledge graph, which is the only data known to the broker. When a specific query is made on the consumer side, the method first composes a forward rewriting from the knowledge graph model to the consumer model, along with a reverse rewriting from the knowledge graph model to the source. This rewriting can be automatically simplified to a direct mapping from the source to the consumer model and, finally, is automatically transformed into optimized database code that directly transports the data from the source to the consumer, without any central platform seeing the data. A few start-ups are just beginning to offer commercial usage of such advanced methods on the market.

7 Projection 6: Industrial GDPR

Many legal and regulatory uncertainties exist regarding data protection in industrial environments. The study indicates disagreement among experts about whether new industry-oriented data protection regulations, similar to the European Union's GDPR, which regulates data protection and privacy within the EU, will emerge. Business-to-business data exchange would be governed at an organization level. However, the probability of such a law is estimated at around 60%.

Driven by concerns about the growing influence of dominant Internet players, the issue of data sovereignty has captured much attention in recent European debates. Data sovereignty is defined as the ability to freely decide about the usage of your own data. For individual persons, it is regulated by the General Data Protection Regulation (GDPR) adopted in 2018 by the European Union and also taken up in many other parts of the world. Aspects of the GDPR are already applicable in production today. For example, connected smart devices on the shop floor may collect personal data if a link to an employee can be established, for instance, through idle times correlating with breaks. Data exchange by increasingly individualized goods can become similarly critical regarding privacy aspects.

Data sovereignty is also an important issue for engineering and production organizations who need to protect the competitive advantage contained in and the knowledge derived from their data. Controlled data sharing and value appropriation thus becomes an essential complement to the value creation from IIoT data integration and analytics. It becomes even more complex in light of networks of production sites. Europe's Digital Single Market Strategy encourages the instantiation of a data-sharing economy, yet the legal schemes are still unclear (Wiebe, 2017).

There are various challenges for regulations concerning data sovereignty in industrial settings. The allocation problem addresses the question of data ownership. For instance, in a connected car equipped with hundreds of sensors, does the data belong to the manufacturer or the owner, and who gets the right to exploit the data economically? The example of the GDPR makes many end users aware of how

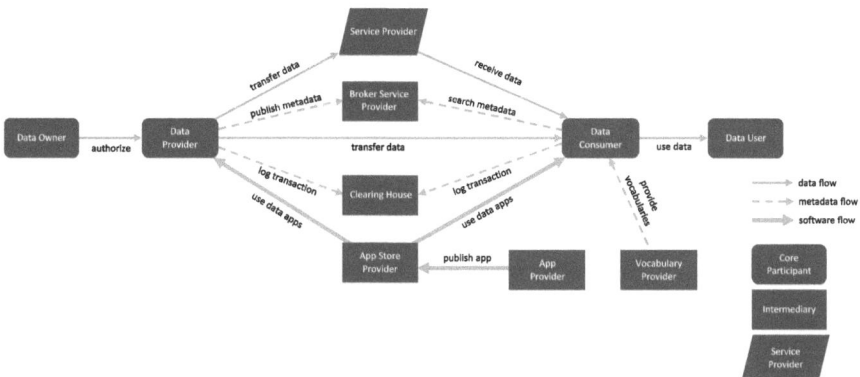

Fig. 2 Rules and interactions in a data space according to the IDS reference architecture 3.0 (Source: Otto et al., 2019)

valuable their data is to Internet corporations that offer free services and sell data for advertising purposes in the background. The specification problem has is based on the fact that data are inherently abstract and are not, for example, subject to the laws on physical property. Data represents a state of information during creation, processing, and storage.

The German Fraunhofer Society initiated the so-called Industrial Data Space initiative in 2014 to analyze corporate data sovereignty needs and to develop a reference architecture for alliance-driven platforms for data ecosystems (Otto & Jarke, 2019). From this early work, the International Data Spaces Association emerged to standardize the principles of self-protected, cross-organizational data spaces that enable trusted smart services, data-based business models, and cross-company data exchange while securing data sovereignty for all participants. In addition to the task of data mediation mentioned in Projection 5, data sovereignty requires agreement on and support for monitoring or enforcement tools concerning not just the data access rights but also permitted or prohibited data usage.

To achieve this goal, existing standards, technologies, and governance models are combined. The International Data Space Reference Architecture Model (IDS-RAM, Otto et al., 2019) specifies how the different technologies and stakeholders should work together and provides a structure for using IIoT data and cross-company IIoT technologies. The IDS-RAM categorizes participating actors into the roles of Core Participants, Intermediary, Software/Service Provider, and Governance Body, as shown in Fig. 2. The data provider, who can also be the data owner, provides data to the data consumer via a service provider. The data consumer provides these data to the data user under well-defined and ideally automatically monitored, or even enforced, usage controls. Clearing houses record all transactions. Broker service providers store and manage metadata about the data sources available in the IDS. The same person or company can take on different roles within the IDS.

An identity provider issues identities to entities in the IDS to ensure secure operation and to prevent unauthorized access to data. Two central technical components are part of the architecture. First, the IDS connector protects the inflow and outflow of data to/from a partner. Second, usage constraints (or policies) govern data exchange policies within the data space (Schuh et al., 2022). Fair usage and value capture are therefore central aspects of a data space with controlled data sharing that can be used to create new business models which use IIoT data as part of a service or as a product.

Building on earlier research on regional mobility brokers, the German Mobility Data Space (Otto & Burmann, 2021) is one of the largest examples of the application of the IDS architecture. It brings together all large vendors of personal mobility offerings (train systems, rental cars/bikes/scooters, regional transport networks, public weather data, etc.) to share data for so far dozens of innovative and cross-vendor mobility offerings, e.g., personalized weather-dependent travel recommendations, together with combination ticketing and real-time travel updates across different vendors and regions. The even larger Catena-X group of companies is preparing a similar value-added data space for the automotive engineering field (Catena-X, 2021).

The GDPR has created several implications for all organizations dealing with personal data, but also for users interacting with these companies, or even with websites. As a result, users are now forced to engage more closely with their rights. Following the European example, California, for example, has introduced a new privacy law. Similarly, an industrial GDPR would impact processes and systems for both data producers and data consumers. Therefore, data protection in industrial environments needs to be embedded in all levels of enterprise architectures, from technical service-level agreements to cross-organizational aspects of business models.

8 Summary

This chapter aimed to develop an understanding of the challenges expected when dealing with Next Generation Manufacturing, addressing this from a governance perspective, namely, by looking at subscription models, digital services, data sharing, central platforms, data mediators, and industrial GDPR.

Optimizing processes and products by collecting usage data can be discussed in the context of subscription models. The likelihood that subscription models will be the new industry standard for production machinery in 2030 was considered relatively high by the experts. This estimation can be justified by these model's characteristics and the resulting benefits for providers and customers. Subscription models are characterized by the provider retaining ownership of the production machine and instead selling the machine's performance for a regular payment. By recording usage data, the provider can, for example, offer supplementary services to increase the machine's output. Further research could be needed to enable companies aiming to introduce a subscription model to identify in a structured way which data from product usage can be used to analyze and, above all, optimize usage. The customer gets full flexibility, along with low initial investment costs. Simultaneously, the experts agreed that the occurrence of the projection would cause a high impact on companies. This can be explained by the extensive structural and organizational adjustments required to offer a subscription model. Therefore, managers in the industry are advised to seek an exchange of experience with other companies that are already successfully offering such a business model when launching the introduction of a subscription model. Regarding the financing of such a business model changeover, a subscription model should initially be introduced for only a few products.

Furthermore, sharing industrial data with other firms will become more important in the near future, as experts consider an increase in the sharing of usage and production data to be highly likely. Organizations that share data with different players are highly likely to obtain competitive advantages over organizations that do not. Experts assessed the impact of data sharing on the competitiveness of firms as high, referring to the shift in knowledge accessibility, resulting in new opportunities and enabling advances in production as well as in supply chains. In addition, data sharing requires new notions of trust and security to be formed and implemented. If these are in place, new business models that enable and govern data exchange on platforms can emerge. A regulatory framework, technical requirements, and the appropriate usage of shared data are prerequisites discussed by individual experts, indicating the complexity of data sharing in firms. Further research is needed to investigate how to overcome these challenges in order to capture value in the context of business models in Next Generation Manufacturing. In particular, future research should consider exploring the circumstances under which firms are willing to share industrial data in an environment of interconnected businesses, with the background of a shift to a more sustainable value chain.

Addressing the expected future of a networked production landscape, efforts are being made worldwide to provide customized offers as digital services, depending on how the customer uses the individual hardware or even the network of production units. There was disagreement among the individual experts of the survey on whether the capabilities and functionality of hardware will shift completely to (digital) services or whether we will instead see (digital) services via intelligent analysis software in combination with or as an extension of the hardware's functionality in order to achieve the desired economic and ecological goals in terms of the worldwide sustainability trend.

Considering the potential shift toward competition in digital services and their importance in the near future, legislative, software-, and hardware-related questions and research opportunities arise regarding the customs clearance of digital goods, licensing possibilities, and open access. As subscription models, digital services, and data sharing gain importance, the question is how and by whom these collaboration modes are governed and orchestrated. Problems related to data leaks and cyber privacy should be addressed to safeguard know-how and prevent production sites in high-wage countries from being jeopardized.

Unlocking and utilizing data in cross-company settings offers many new opportunities for companies in Next Generation Manufacturing settings. To implement the exchange of abstract data products, however, framework conditions like interfaces and usage policies need to be defined. The survey shows disagreement among the respondents as to whether these will be applied individually or enforced externally by a data mediator. Existing approaches like the IDS clearinghouse would support data policies such as usage controls.

Consumer markets suggest that mono- and oligopolistic market structures where central platforms orchestrate the relationships may be the future. However, in industrial settings, no major platform has been able to establish itself as the dominant leader. Alliance-driven approaches under shared leadership seem more likely. Ongoing initiatives, such as IDS and GAIA-X, provide potential solutions. This study can also be considered an interesting contribution to the European GAIA-X debate (https://www.gaia-x.eu). GAIA-X wants to address challenges regarding data sovereignty at three levels: the lack of high performance and secure network, storage, and computing infrastructure ecosystems ("hyperscaler") in Europe; the need for federation services to enable sovereign and privacy-preserving data sharing and mediation; and the promotion of cross-organizational value creation and appropriation through a data ecosystem of data-centric services and new business models. This study contributes valuable expert insights on many of these strategic questions, focusing on the interplay between the data and the business levels.

Driven by concerns about the growing influence of dominant Internet players, the GDPR was adopted in 2018 by the European Union to regulate data protection for individual persons. However, many legal and regulatory uncertainties exist for industrial data sharing scenarios. The survey indicates disagreement among the experts on whether new industry-oriented data protection regulations similar to the GDPR will emerge. The International Data Spaces initiative offers a reference architecture that enables such regulations to be embedded within enterprise architectures. Managers face the risks of building a platform (e.g., Yoffie et al., 2019), but following an alliance-driven approach where the platform (or multiple platforms) is operated under shared leadership could provide an effective alternative. Finally, the protection of shared personal data in organizations needs to follow the GDPR, impacting internal processes and systems for both data producers and customers.

Acknowledgment Funded by the Deutsche Forschungsgemeinschaft (DFG, German Research Foundation) under Germany's Excellence Strategy – EXC-2023 Internet of Production – 390621612.

References

Barney, J. (1991). Firm resources and sustained competitive advantage. *Journal of Management, 17*(1), 99–120. https://doi.org/gpm.
Becker, F., Bibow, P., Dalibor, M., Gannouni, A., Hahn, V., Hopmann, C., ... Wortmann, A. (2021). A conceptual model for digital shadows in industry and its application. In *International Conference on Conceptual Modeling* (pp. 271–281). Springer. https://doi.org/hg45.

Boudreau, K. (2010). Open platform strategies and innovation: Granting access vs. devolving control. *Management Science, 56*(10), 1849–1872. https://doi.org/cp9trp.

Boudreau, K. J., & Jeppesen, L. B. (2015). Unpaid crowd complementors, the platform network effect mirage. *Strategic Management Journal, 36*, 1761–1777. https://doi.org/gc8sv6.

Braud, A., Fromentoux, G., Radier, B., & Le Grand, O. (2021). The road to European digital sovereignty with Gaia-X and IDSA. *IEEE Network, 35*(2), 4–5. https://doi.org/hg46.

Brecher, C., Butz, F., Kaever, M., Nagel, E., Neus, S., Rettich, T., Steinert, A., Welling, D., Wiesch, M., Zeis, M., & Zeppenfeld, C. (2021). Nachhaltige Geschäftsmodelle für Werkzeugmaschinen. In *Internet of production—Turning data into sustainability* (pp. 433–462). Fraunhofer-Gesellschaft.

Brillowski, F., Dammers, H., Koch, H., Müller, K., Reinsch, L., & Greb, C. (2021). Know-how transfer and production support systems to cultivate the internet of production within the textile industry. In *International conference on intelligent human systems integration* (pp. 309–315). Springer. https://doi.org/hg47.

Burmeister, C., Lüttgens, D., & Piller, F. T. (2016). Business model innovation for industrie 4.0: Why the 'industrial internet' mandates a new perspective on innovation. *Die Unternehmung, 2*, 124–152. https://doi.org/gh79t7.

Catena-X (2021). Catena-X automotive network—the gateway to a digital economy. https://catena-x.net/en/

Cennamo, C., & Santalo, J. (2013). Platform competition: Strategic trade-offs in platform markets. *Strategic Management Journal, 34*(11), 1331–1350. https://doi.org/gc8svs.

Chekurov, S., Metsä-Kortelainen, S., Salmi, M., Roda, I., & Jussila, A. (2018). The perceived value of additively manufactured digital spare parts in industry: An empirical investigation. *International Journal of Production Economics, 205*, 87–97. https://doi.org/gfnbww.

Clark, T. H., & Hammond, J. H. (1997). Reengineering channel reordering processes to improve total supply-chain performance. *Production and Operations Management, 6*(3), 248–265. https://doi.org/dp2tfv.

Dattée, B., Alexy, O., & Autio, E. (2018). Maneuvering in poor visibility: How firms play the ecosystem game when uncertainty is high. *Academy of Management Journal, 61*(2), 466–498. https://doi.org/gdsh4z.

Deloitte. (2014). New business models with data. https://ec.europa.eu/futurium/sites/futurium/files/deloitte_pov_new_business_models_with_data.pdf

Eisenmann, T., Parker, G., & Van Alstyne, M. W. (2006). Strategies for two-sided markets. *Harvard Business Review, 84*(10), 92.

Fitzgerald, M. (2013). An internet for manufacturing. MIT Technology Review.

Gassmann, O., Frankenberger, K., & Csik, M. (2013). Geschäftsmodelle aktiv innovieren. In *Das unternehmerische Unternehmen* (pp. 23–41). Springer Gabler.

Gawer, A. (2014). Bridging differing perspectives on technological platforms: Toward an integrative framework. *Research Policy, 43*(7), 1239–1249. https://doi.org/gc8sc5.

Gleim, L., Pennekamp, J., Liebenberg, M., Buchsbaum, M., Niemietz, P., Knape, S., ... Wehrle, K. (2020). FactDAG: Formalizing data interoperability in an internet of production. *IEEE Internet of Things Journal, 7*(4), 3243–3253. https://doi.org/ggqvdf.

Gregory, R. W., Henfridsson, O., Kaganer, E., & Kyriakou, H. (2021). The role of artificial intelligence and data network effects for creating user value. *Academy of Management Review, 46*(3), 534–551. https://doi.org/ggnjt7.

Huttunen, H., Seppala, T., Lahteenmaki, I., & Mattila, J. (2019). What are the benefits of data sharing? Uniting supply chain and platform economy perspectives. Uniting Supply Chain and Platform Economy Perspectives.

Iansiti, M., & Lakhani, K. (2020). Competing in the age of AI. *Harvard Business Review, 98*(1), 60–67.

Jacobides, M. G., Cennamo, C., & Gawer, A. (2018). Towards a theory of ecosystems. *Strategic Management Journal, 39*(8), 2255–2276. https://doi.org/gd845g.

Jarke, M., Jelassi, M. T., & Shakun, M. F. (1987). MEDIATOR: Towards a negotiation support system. *European Journal of Operational Research, 31*(3), 314–334. https://doi.org/fxgfx7.

Jarke, M., Lenzerini, M., Vassiliou, Y., & Vassiliadis, P. (2002). *Fundamentals of data warehouses.* Springer Science & Business Media.

Kopalle, P. K., Kumar, V., & Subramaniam, M. (2020). How legacy firms can embrace the digital ecosystem via digital customer orientation. *Journal of the Academy of Marketing Science, 48*(1), 114–131. https://doi.org/gj27r5.

Korbel, J. J., Blankenhagel, K. J., & Zarnekow, R. (2019). The role of the virtual asset in the distribution of goods and products. In 25th Americas Conference on Information Systems. Cancun, Mexico.

Lenzerini, M. (2019). Direct and reverse rewriting in data interoperability. In *International conference on advanced information systems engineering* (pp. 3–13). Springer. https://doi.org/hg5c.

Levitin, A. V., & Redman, T. C. (1998). Data as a resource: Properties, implications, and prescriptions. *MIT Sloan Management Review, 40*(1), 89.

Li, Z. X., & Agarwal, A. (2017). Platform integration and demand spillovers in complementary markets: Evidence from Facebook's integration of Instagram. *Management Science, 63,* 3438–3458. https://doi.org/gc8svg.

McCarthy, D. M., Fader, P. S., & Hardie, B. G. (2017). Valuing subscription-based businesses using publicly disclosed customer data. *Journal of Marketing, 81*(1), 17–35. https://doi.org/gfw27t.

Meier, H., & Uhlmann, E. (Eds.). (2012). *Integrierte Industrielle Sach- und Dienstleistungen: Vermarktung, Entwicklung und Erbringung hybrider Leistungsbündel.* Springer Science & Business Media.

Mori, D. M. G. (2021). DMG Mori launches subscription business with PAYZR. https://en.dmgmori-ag.com/resource/blob/564672/7138e2982e5897593ce7b68a18f82781/dmg-mori-launches-subscription-business-with-payzr-pdf-data.pdf.

Otto, B., & Burmann, A. (2021). Europäische Dateninfrastrukturen—Ansätze und Werkzeuge zur Nutzung von Daten zum Wohl von Individuum und Gesellschaft. *Informatik Spektrum, 44,* 283–291. https://doi.org/hg5d.

Otto, B., & Jarke, M. (2019). Designing a multi-sided data platform: Findings from the international data spaces case. *Electronic Markets, 29*(4), 561–580. https://doi.org/ggqvq9.

Otto, B., Steinbuss, S., Teuscher, A., et al. (2019). *IDS Reference Architecture Model Version 3.0.* International Data Spaces Association.

Parker, G. G., & Van Alstyne, M. W. (2018). Innovation, openness, and platform control. *Management Science, 64*(7), 3015–3032. https://doi.org/gc8svd.

Parker, G. G., Van Alstyne, M. W., & Choudary, S. P. (2016). *Platform revolution: How networked markets are transforming the economy and how to make them work for you.* WW Norton & Company.

Piller, F. T., Van Dyck, M., Lüttgens, D., & Diener, K. (2021). Positioning strategies in emerging industrial ecosystems for industry 4.0. In Proceedings of the 54th Hawaii International Conference on System Sciences (p. 6153). https://doi.org/hgz3.

Porter, M. E., & Heppelmann, J. E. (2015). How smart, connected products are transforming companies. *Harvard Business Review, 93*(10), 96–114.

Rappa, M. A. (2004). The utility business model and the future of computing services. *IBM Systems Journal, 43*(1), 32–42. https://doi.org/bpm854.

Riesener, M., Doelle, C., Ebi, M., & Perau, S. (2020). Methodology for the implementation of subscription models in machinery and plant engineering. *Procedia CIRP, 90,* 730–735. https://doi.org/hg5f.

Rolls Royce. (2012). Rolls-Royce celebrates 50th anniversary of Power-by-the-Hour. https://www.rolls-royce.com/media/press-releases-archive/yr-2012/121030-the-hour.aspx

Rudolph, T., Bischof, S. F., Böttger, T., & Weiler, N. (2017). Disruption at the door: A taxonomy on subscription models in retailing. *Marketing Review St. Gallen, 5,* 18–25.

Schuh, G., Frank, J., Jussen, P., Rix, C., & Harland, T. (2019). Monetizing industry 4.0: Design principles for subscription business in the manufacturing industry. In 2019 IEEE international conference on engineering, technology and innovation (ICE/ITMC) (pp. 1–9). https://doi.org/ghnw75.

Schuh, G., Jarke, M., Gützlaff, A., Koren, I., Janke, T., & Neumann, H. (2022). Review of commercial and open technologies available for industrial internet of things. In D. Mourtzis (Ed.), *Design and operation of production networks for mass personalization in the era of cloud technology* (pp. 209–241). Elsevier.

Schuh, G., Riesener, M., Prote, J. P., Dölle, C., Molitor, M., Schloesser, S., ... Tittel, J. (2020). Industrie 4.0: Agile Entwicklung und Produktion im Internet of Production. In *Handbuch Industrie 4.0: Recht, Technik, Gesellschaft* (pp. 467–488). Springer.

Schuh, G., Salmen, M., Jussen, P., Riesener, M., Zeller, V., & Hensen, T. (2017). Geschäftsmodell-Innovation. In G. Reinhart (Ed.), *Handbuch Industrie 4.0: Geschäftsmodelle, Prozesse, Technik* (pp. 3–29). Hanser.

Stefansson, G. (2002). Business-to-business data sharing: A source for integration of supply chains. *International Journal of Production Economics, 75*(1–2), 135–146. https://doi.org/c8nkmj.

Tukker, A. (2004). Eight types of products—Service system: Eight ways to sustainability? Experiences from SusProNet. *Business Strategy and the Environment, 13*(4), 246–260. https://doi.org/bpr234.

Tzuo, T. (2018). *Subscribed: Why the subscription model will be your company's future—And what to do about it.* Portfolio/Penguin.

Warrillow, J. (2015). *The automatic customer: Creating a subscription business in any industry.* Penguin.

West, J. (2003). How open is open enough?: Melding proprietary and open source platform strategies. *Research Policy, 32*(7), 1259–1285. https://doi.org/cf6dhc.

Wiebe, A. (2017). Protection of industrial data—A new property right for the digital economy? *Journal of Intellectual Property Law & Practice, 12*(1), 62–71. https://doi.org/ggf2sf.

Wiederhold, G. (1992). Mediators in the architecture of future information systems. *Computer, 25*(3), 38–49. https://doi.org/fvzz7f.

Yoffie, D. B., Gawer, A., & Cusumano, M. A. (2019). *A study of more than 250 platforms reveals why most fail.* University of Surrey.

Zhao, Y., Von Delft, S., Morgan-Thomas, A., & Buck, T. (2020). The evolution of platform business models: Exploring competitive battles in the world of platforms. *Long Range Planning, 53*(4), 101892. https://doi.org/ggmt3p.

Zhu, F., & Iansiti, M. (2019). Why some platforms thrive and others don't. *Harvard Business Review, 62*(1), 118–125.

Organization Routines in Next Generation Manufacturing

Philipp Brauner, Luisa Vervier, Florian Brillowski, Hannah Dammers, Linda Steuer-Dankert, Sebastian Schneider, Ralph Baier, Martina Ziefle, Thomas Gries, Carmen Leicht-Scholten, Alexander Mertens, and Saskia K. Nagel

Abstract Next Generation Manufacturing promises significant improvements in performance, productivity, and value creation. In addition to the desired and projected improvements regarding the planning, production, and usage cycles of products, this digital transformation will have a huge impact on work, workers, and workplace design. Given the high uncertainty in the likelihood of occurrence and the technical, economic, and societal impacts of these changes, we conducted a technology foresight study, in the form of a real-time Delphi analysis, to derive reliable future scenarios featuring the next generation of manufacturing systems. This chapter presents the organization dimension and describes each projection in detail, offering current case study examples and discussing related research, as well as implications for policy makers and firms. Specifically, we highlight seven areas in which the digital transformation of production will change how we work, how we organize the work within a company, how we evaluate these changes, and how

P. Brauner (✉) · L. Vervier · M. Ziefle
Human-Computer Interaction Center, RWTH Aachen University, Aachen, Germany
e-mail: brauner@comm.rwth-aachen.de; vervier@comm.rwth-aachen.de; ziefle@comm.rwth-aachen.de

F. Brillowski · H. Dammers · T. Gries
Institut für Textiltechnik, RWTH Aachen University, Aachen, Germany
e-mail: florian.brillowski@ita.rwth-aachen.de; hannah.dammers@ita.rwth-aachen.de; thomas.gries@ita.rwth-aachen.de

L. Steuer-Dankert · S. Schneider · C. Leicht-Scholten
Research Group Gender and Diversity in Engineering, RWTH Aachen University, Aachen, Germany
e-mail: linda.steuer@gdi.rwth-aachen.de; sebastian.schneider@gdi.rwth-aachen.de; carmen.leicht@gdi.rwth-aachen.de

R. Baier · A. Mertens
Institute of Industrial Engineering and Ergonomics, RWTH Aachen University, Aachen, Germany
e-mail: r.baier@iaw.rwth-aachen.de; a.mertens@iaw.rwth-aachen.de

S. K. Nagel
Human Technology Center/Applied Ethics, RWTH Aachen University, Aachen, Germany
e-mail: saskia.nagel@humtec.rwth-aachen.de

F. T. Piller et al. (eds.), *Forecasting Next Generation Manufacturing*, Contributions to Management Science, https://doi.org/10.1007/978-3-031-07734-0_5

employment and labor rights will be affected across company boundaries. The experts are unsure whether the use of collaborative robots in factories will replace traditional robots by 2030. They believe that the use of hybrid intelligence will supplement human decision-making processes in production environments. Furthermore, they predict that artificial intelligence will lead to changes in management processes, leadership, and the elimination of hierarchies. However, to ensure that social and normative aspects are incorporated into the AI algorithms, restricting measurement of individual performance will be necessary. Additionally, AI-based decision support can significantly contribute toward new, socially accepted modes of leadership. Finally, the experts believe that there will be a reduction in the workforce by the year 2030.

[Abstract generated by machine intelligence with GPT-3. No human intelligence applied.]

1   Introduction

The (first) Industrial Revolution had an enormous impact on the world of work and society: instead of being carried out in the workers' homes or in craft shops, production was increasingly shifted to factories. This made manufacturing cheaper and more productive, but it also had an enormous impact on society and led to social grievances (Engels, 1971). The upcoming digital transformation of production termed Industry 4.0 or the Industrial Internet of Things promises significant improvements in performance, productivity, and value creation (Kagermann, 2015; Liao et al., 2017; Brauner et al., 2022). However, in addition to the desired and projected improvements regarding the planning, production, and usage cycles of products, this digital transformation will have a huge impact on work, workers, and workplace design (Acemoglu & Restrepo, 2017; Brynjolfsson & Mitchell, 2017). In this chapter, we will therefore highlight seven areas in which the digital transformation of production will change how we work, how we organize the work within a company, how we evaluate these changes, and how employment and labor rights will be affected across company boundaries. Thus, we here consider what implications Next Generation Manufacturing will have for work, the workforce, and society.

In detail, the proposed concepts will include changes in work organization and work structures inside and outside companies. Improved sensors and actuators and faster, smarter control systems mean that automation and robots will take on more tasks and work more closely with humans. New forms of human-robot collaboration will also emerge (Villani et al., 2018; Borenstein, 2011). At the same time, production data within companies and along supply chains will be recorded more accurately. Improved data models can then be analyzed by advanced AI methods and used either for full automation or to support human decision-makers. Therefore, a new form of hybrid intelligence—the combination of artificial and human intelligence—will support the production of, decision-making around, and creation of new products. Also, models and digital shadows can be generated of not only production itself but also the employees in the sociotechnical system. While this can be used for better coordination between workers and the production system, it can also be used for personnel management or employee control. It is therefore to be expected that this digital transformation will also affect the workforce, employee rights, and employment (Autor, 2015; Lepore, 2019).

Using a novel real-time Delphi approach (see chapter "Applying the Real-Time Delphi Method to Next Generation Manufacturing" for a presentation of the method and the sample, as well as chapter "Big Picture of Next Generation Manufacturing" for an overview of the results), we developed propositions for different scenarios within Next Generation Manufacturing in 2030. As suggested by Gawer (2014), we used an integrative framework for platforms, distinguishing four dimensions: governance (e.g., open forms of collaboration; see chapter "Governance Structures in Next Generation Manufacturing"), organization (e.g., boundaries and decision-making; see this chapter), capabilities (e.g., hybrid intelligence; see chapter "Capability Configuration in Next Generation Manufacturing"), and interfaces (e.g., open APIs and human-machine interfaces; see chapter "Interface Design in Next Generation Manufacturing"). In addition, and influenced by our shared experiences during the COVID-19 pandemic, we added a fifth cluster of propositions addressing the need for resilience in future digital manufacturing (see chapter "Resilience Drivers in Next Generation Manufacturing"). We provide a set of 24 validated projections based on 1.930 quantitative estimations and 629 qualitative arguments from 35 industrial and academic experts from Europe, North America, and Asia. In this way, we provide a foundation upon which academic discussion can be grounded and which can support decision making regarding future technological developments and economic impacts that go beyond current speculation and isolated research.

In this chapter, we analyze whether production will profit from the increased autonomy of robots (P7). We further explore whether collaborations between human and artificial intelligence in "hybrid intelligence" can be expected to be meaningful in producing companies (P8) and whether AI-based assistants will reshape decision-making (P9). We also address the question of whether the digital transformation will change work culture and leadership (P10) and what opportunities and challenges will emerge if digital shadows of employees are created (P11). Finally, we look at how the changes triggered by Next Generation Manufacturing will affect workers' rights (P12) and what effects this will have on the labor market (P13) (see Fig. 1).

2 Projection 7: Autonomous Robots

It is impossible to imagine the large-scale production of the last 50 years without robots, which have traditionally been used for the full automation of processes. Prominent robotic applications of this period include, for example, spot welding, spray painting, assembly, machining, and electronic testing. The tasks included in these applications can be performed very well by robots due to their positive

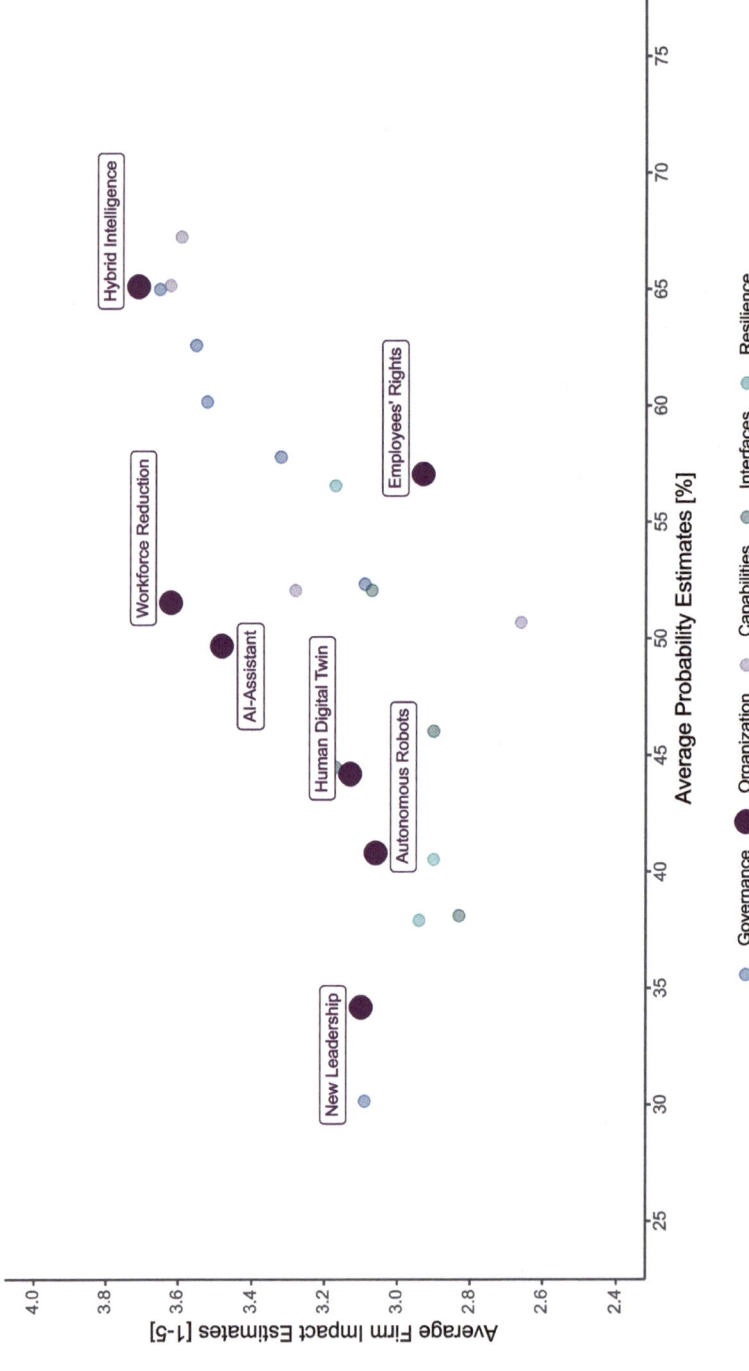

Fig. 1 Expert assessment of organization routines in Next Generation Manufacturing (see chapter "Big Picture of Next Generation Manufacturing" for the full results of the Delphi survey)

characteristics regarding reliability, durability, accuracy, and precision (Grau et al., 2017).

However, traditional robots are not suitable for fulfilling every task. For example, they still fail during unstructured tasks, when reacting to uncertainties or frequent changes, or when their tasks require cognitive or motoric flexibility. In recent years, production companies have replaced traditional industrial robots with so-called collaborative robots (or *cobots* for short) that are equipped with additional sensor and software systems to enable human-robot or robot-robot collaboration (Villani et al., 2018). Due to their ability to perceive people and obstacles, costly and impedimentary elaborate safety measures such as protective fences and light barriers can be removed. In addition, due to simple teaching methods, usage of cobots enables higher flexibility and partial automation of work processes. Other examples include automated guided vehicles (AGVs) that can move autonomously through shop floors and can thus further increase the flexibility and autonomy of robots (Castro et al., 2021).

In the survey, the experts reported a low average probability that collaborative robots which move autonomously through the shop floor will have replaced most conventional robots in protective cells by 2030 ($M = 40.81\%$, $SD = 25.72\%$). The experts consider 2030 to be too early for such a full transformation on the shop floor because of the existing technical and safety regulations. In particular, high investment costs are seen as a major hurdle, so it is expected that a considerable number of cheap and efficient robots that do not need to interact with humans or move freely will remain. For high-volume production of many product groups, where high speed in fixed setups is needed, traditional robots still have advantages regarding lower cycle times, lower cost, and higher process capability. However, the experts mentioned that significant progress has been made in special applications where human-robot collaboration and autonomously moving robots are feasible and economically viable. There was low agreement between the experts on our panel on this topic ($IQR = 3.5$). As some experts believe this projection to be more likely, it should be investigated whether these differences are due to different application domains or industry sectors.

If this projection comes true, the experts estimate that it will have a moderate average firm impact of $M = 3.06$. Some experts assume there will be significant improvements in flexibility in certain use cases so that a fast reaction to changes in demand, changing technologies, and new products is possible. According to the experts, this will also lead to a significant impact on the attainable productivity and return on investment (ROI). Other experts see a high level of collaborative robots being realized on shop floors already today, leading to a lower impact similar to ordinary efficiency improvements.

In conclusion, to transform conventional robotic production scenarios into robots moving freely on the shop floor, interacting with workers, and autonomously making decisions, technologies and AI methods must be (further) developed and costs reduced. A successful conversion can only proceed if human factors such as trust, acceptance, and ergonomics are embraced (Brauner et al., 2022), while representatives such as trade unions and work councils must be involved at an early stage. In addition, both safety and security aspects must be considered. It is essential to ensure that human workers are safe and that data leaks and hacks are prevented. Furthermore, ethical and legal restrictions must also be considered, for example, regarding responsibility, warranty, and liability.

To advance the use of autonomous robots in production, specific applications must be identified. Among these are, for example, more complex tasks in logistics for material transport between different workstations by AGVs (Gualtieri et al., 2021). Other applications for collaborative robots can primarily be found in areas detrimental to humans, e.g., within the handling of chemicals (automotive paint shop, Ford) and ergonomically unfavorable positions (insertion of rivets inside aircraft fuselage, Airbus) or in repetitive assembly (filigree micro parts, Albrecht Jung). In addition, it has recently been shown that collaborative robots and human-robot collaboration can be used to respond flexibly to needs during the COVID-19 pandemic (see also (P24) in chapter "Resilience Drivers in Next Generation Manufacturing"), e.g., the ramping up of propeller production while maintaining social distancing (Malik et al., 2021).

3   Projection 8: Hybrid Intelligence

Monitoring new forms of data in production, as well as significant advances in process mining methods, leads to productive applications of AI in so-called hybrid intelligence, i.e., supplementing human decision-making processes with an AI system. Hybrid intelligence can be applied at all levels, from production on the shop floor to strategic management decision-making.

A potential use case of hybrid intelligence is given by the vision of control of global production streaming from the IoP. This approach aims to improve decision quality and speed in dynamic production environments based on the targeted collection of machine and sensor data. The corresponding data evaluations are intended to be hybrid decisions made using AI and expert knowledge. Such a hybrid intelligence approach is promising for strategic, tactical, and operational levels. By feeding the information back into the data sources, e.g., the IT systems that control the machines, optimizations could be realized in real time throughout an entire production network (WGP, 2019; Schuh et al., 2019).

The experts predict that hybrid decision-making in the form of human-AI collaboration will have a strong impact on the management profession in the future and is very likely to occur. They predict hybrid intelligence will occur with an average expected likelihood of 65.13% ($SD = 23.66\%$) and will have a high impact on the industry ($M = 3.71$). There was little deviation across experts from industry and academia.

However, the experts interviewed in this Delphi study did not indicate a clear position on the projection that says that in 2030, strategic production decisions will be executed through close interaction between humans and AI-based algorithms. Some of the experts assume that human decisions will be supplemented by AI in the future, of whom some remark that this inclusion of AI is already happening today, e.g., in finance. Others hold that this trend will take more time and that human decisions in everyday work will always be influenced by their own affective feelings rather than by AI.

In order to use hybrid intelligence sensibly in the production of the future, further understanding of the conditions is required for its implementation. Hybrid intelligence is a form of human-machine interaction, raising questions about the reliability and fairness of such systems and, consequently, about the attribution of responsibility in the context of shared tasks.

First, working in cooperation with an accurate and precise AI system potentially induces the phenomenon of social loafing as workers adapt to human-machine collaborations, i.e., human experts may show a tendency to go to less effort when working together with an algorithm than when working alone (Parasuraman & Manzey, 2010; Brauner et al., 2019). Instead of forming decisions supplemented by the input of an AI system, the human expert may tend to rely on the decision of the system. Thus, overreliance could lead to the human expert not sufficiently engaging in the decision-making, ultimately leading to complex questions of responsibility attribution (Zerilli et al., 2019; Liehner et al., 2021). If the human expert does not function at least as a co-decision-maker, effectively, the machine is taking autonomous decisions. However, we generally do not hold machines responsible, and the lack of suitable agents for responsibility attribution gives rise to further moral issues, such as on autonomy and contestability.

Second, the past has shown that AI systems are not self-evidently fair, but that biased training data may reinforce societal biases and algorithmic design may create discriminatory outcomes. Yet high-level decision-making in an industrial context involves decisions of importance to, e.g., workforces in the industry. Therefore, a differentiated discussion on the fairness of AI systems is crucial to prevent the emergence of avoidable distributive injustices (Binns, 2018).

On the upside, hybrid intelligence offers significant opportunities for improving high-level decision-making. From applications of decision support systems in medicine, there is evidence that working with the support of an automated system may, first, free up cognitive load, which is necessary for more complex cognitive tasks such as those requiring abstract thinking, and, second, improve decision-making processes as a direct result of technological integration (Patel et al., 2002).

4   Projection 9: AI-Based Assistants

Making a thorough decision is a time-consuming process that requires a lot of different information. The required information must be gathered and integrated from various sources and must then be evaluated according to various criteria. Based on this evaluation, a decision-maker can conclusively make a well-founded decision.

In decision theory, models often assume an ideal rational being and decision process. Yet decision-making is often based on experience, implicit knowledge, and emotions (Lerner et al., 2015), and people are easily distracted by complacency or bad user interfaces (Brauner et al., 2019). In the past, automated model-based approaches have therefore been developed to objectify decision-making and avoid an intuitive procedure. Especially when an excessive amount of information must be considered, algorithms (e.g., for scheduling) are preferably utilized. However, in the context of safety-critical decisions, algorithms are only used as a decision support due to legal and ethical implications. The decision authority in these cases remains with the human. In contrast, supporting processes already rely heavily on different types of AI (e.g., decision trees or neuronal networks), and the decision authority is shifted toward algorithms. Nevertheless, unwanted decisions can still occur and lead to severe consequence, like Dow Jones' "Flash Crash" in 2010 which caused a market value loss of one trillion dollars within a few minutes (Subrahmanyam, 2013). Furthermore, algorithms tend to be discriminating when trained with a biased dataset, e.g., in job application processes or for chatbots (Hacker, 2018; Neff & Nagy, 2016). Considering the opportunities and dangers of automated decision-making, demand for third-party auditing of algorithms is rising (Wachter et al., 2017).

Most of the experts share these sentiments and raised concerns regarding safety-critical and discriminating decisions and thus still expect people to be the central decision-making figures in the future, contributing with their experience and knowledge. According to the experts, humans will still possess the decision authority and will only be supported by AI systems. Due to this AI support, humans will be able to make quicker decisions, especially given the ever-increasing decision complexity and data availability.

Overall, experts from our panel anticipate a rather high impact ($M = 3.48$) if the projection becomes a reality but are divided about how probable this is ($M = 49.68\%$, $SD = 22.61\%$). There was a medium level of disagreement ($IQR = 3.50$) among the experts when confronted with the projection that in 2030 operative production decisions will no longer lie with people as they will be made by AI-based decision-making agents. However, all the experts acknowledged that AI assistance offers great opportunities for cost and time savings, potentially resulting in a disruptive change in decision-making processes.

Suitable use cases of AI assistance are manifold across various industries and domains. In the context of the IoP, production environments like machining manufacturing or process planning of high-performance fiber-reinforced composites (Brillowski et al., 2020; Schemmer et al., 2020; Brillowski et al., 2021) are the subjects of research. In both cases, a decision-supporting tool ("smart expert") is trained with available and task-specific data (digital shadow). To achieve the best possible usability, the tools must be designed to foster trust, comprehensibility, and availability of crucial information. An AI-based decision tool consists of autonomous decision-making algorithms which either possess the final authority over a decision (machining production) or have an advisory function (planning), only providing suggestions so that the decision authority remains with the human. Overall, applying AI assistants can improve efficiency within production environments significantly and contribute toward broader acceptance of autonomous agents.

While novel access to data and data processing technologies enables more efficient decision processes, there is an inherent risk of automation bias. Humans rely heavily on automatically generated suggestions and simultaneously neglect to scrutinize them critically (Cummings, 2017; Skitka et al., 1999; Brauner et al., 2019). Furthermore, there are ethical and legal implications regarding the AI's reliability and ultimately also its accountability, especially in the case of a fatal decision. A need for a user-centered and participatory design thus arises to achieve explainability and comprehensibility of "black-box" AI models. Only by incorporating the user appropriately is joint decision-making by AI assistants and humans possible. In such a scenario, only outliers and crucial, safety-critical decisions will be made by humans, leaving support process to an autonomous decision agent. Within the Delphi study conducted, the experts agreed that this is the most probable AI assistance scenario within the medium-term future.

5   Projection 10: New Leadership

Artificial intelligence, machine learning, and data infrastructure are changing the way decisions are made by people and organizations. The term "algorithmic management" describes partially or fully automated decision-making processes that were previously carried out by human managers (Jarrahi et al., 2021). For example, AI is already used in recruiting processes, with the aim of making application processes more transparent and allowing firms to carry out selection procedures in a more non-discriminatory way (Tambe et al., 2019; Ochmann & Laumer, 2020). Furthermore, algorithmic management is used in crowdworking (Neuburger & Fiedler, 2020; Rani & Furrer, 2021) or platform work (e.g., Uber, Lyft, Deliveroo) (Duggan et al., 2020), but also for work scheduling (Parker & Grote, 2020). However, besides the hoped-for benefits of artificial intelligence in management processes, studies already point to negative effects of automated algorithms. Examples of this are a higher pressure on workers to perform, less control and influence over individual work tasks, lower perceived autonomy, lower morale, and decreased job satisfaction (Kellogg et al., 2020; Parker & Grote, 2020).

In addition to the effects mentioned above, initial studies are also already indicating discriminatory effects of artificial intelligence, particularly in human resource issues. The cause of this is often found in the databases used for artificial intelligence training processes, which are created by humans and already exhibit a bias (Tambe et al., 2019; Todolí-Signes, 2019). Nevertheless, AI promises the potential to support data-intense decisions and pave the way for novel leadership. However, the negative effects of purely AI-based management processes demonstrated in several recent studies (e.g., Tambe et al., 2019; Ochmann & Laumer, 2020) suggest a need for human interaction in leadership. Additionally, they indicate that if AI systems are implemented within leadership processes, these AI-based management systems must incorporate social and normative aspects via their technological structure and, also, via socially sensitive creation processes during the development and testing of the given technology. Thus, in the future, people will and must continue to be a component of management processes, at least in the sense of affecting the AI technologies used in a way that helps to counteract or even to overcome their negative aspects.

Consequently, the projection "new leadership" raises the question of whether increasing implementation of artificial intelligence will also lead to changes in management processes, leadership, and the elimination of hierarchies. The expert panel gave a low probability of 34.19% ($SD = 20.25\%$) and a medium firm impact of $M = 3.10$ for this projection. This low likelihood results either from an adherence to established structures and systems or from an adherence to power structures which might result from a target group bias. However, the experts' assessment was not unanimous ($IQR = 3.00$), and some experts reported a higher probability for the emergence of new leadership through AI-based decision systems than others.

But interaction with artificial intelligence does result in the need to rethink leadership and to develop new forms of leadership styles. By employing AI in leadership decisions, managers can focus on interhuman factors, while the AI supports mainly data-driven reasoning. However, as introduced above, there are many barriers that currently impede a thorough usage of AI, starting with the overall acceptance of AI reasoning. Furthermore, responsibilities for AI-based decisions must be clearly assigned, and social interactions between humans must be captured in AI algorithms. To guarantee that these social interactions are represented in AI algorithms, several strategies within the wider developmental process of such software can be introduced. As an example, the Institut Montaigne (2020) recommends (a) the deployment of good practices that help to prevent the spread of algorithmic biases, e.g., guaranteeing diversity within development teams or creating internal charters that support the developmental processes, (b) special training for technicians and engineers that strengthens their understanding and awareness of algorithmic biases, (c) the introduction of stronger and more sensitive testing practices for algorithms (even including public or "real" settings in which the AI software is to be used), (d) the introduction of a fairness approach that guides the developmental processes and the persons involved in it, and, along with that, (e) the introduction of a stringent list of requirements that the AI software needs to encompass (or not) in order to accurately capture social aspects of interactions between humans and AI algorithms (Institut Montaigne, 2020). Such strategies might be even more important, as there is currently, e.g., no approach to representing intuitive decisions with AI and the consideration of individual framework conditions.

All of this means that only by guaranteeing that social and normative aspects are incorporated into the AI algorithms and by reducing individual performance measurement and, instead, generating AI-compatible KPIs that help to capture employees' perceptions can AI-based decision support significantly contribute toward socially accepted novel leadership.

6   Projection 11: Human Digital Twin

The term "digital twin" has been around for over a decade: Grieves and Vickers (2017) state that the concept was first introduced at a presentation in 2002 and thereafter went by various names until the term digital twin was published by Grieves in 2011. But it is only in the last few years that the idea has received great attention in industry and academia. The core idea is that there is a virtual entity in addition to the physical entity. These two elements are linked within a cycle: data is fed from the real to the virtual domain, and information and processes pass the reverse way. The virtual representation means that simulations, data analysis, etc. are now possible. The close interweaving, often called synchronization, of the real domain with the digital domain allows a continuous optimization process (Jones et al., 2020; Schuh et al., 2019).

It should be explicitly noted that the mere collection of data on work persons and their work performance does not constitute a digital twin. Thus, the digital twin is still a concept that has not yet been implemented. In the following, visions of how the implementation and use of digital twins could be designed are presented. A digital twin of a human could benefit production planning in a variety of use cases on different organizational levels (Mertens et al., 2021). In human-robot collaboration, using knowledge about an employee's experience, handedness, and working methods would enable situation-specific matching of a robot's interaction with its user. For decision tasks, workers could benefit from decision support systems that tailor their interface and the allocation of tasks based on the user's current mental workload. Furthermore, digital twins of employees could enable health strategies in human resource management that are specifically tuned to an individual's needs.

This paragraph reports the results of the study. The experts assessed the probability that, by 2030, a full digital twin of each production worker and all of her or his operations will be used as a tool for production planning and optimization as rather low. The experts from industry estimate the probability to be 47.38%, much more likely than the experts from academia at 37.50%. German experts estimate the probability to be 41.74%, noticeably lower than the experts from the rest of the world (ROW) with 51.25%. There is disagreement among the German experts, but not among the ROW experts. Those who estimate the probability of occurrence to be low identify two central causes: resistance from work councils and unions and legal reasons, in particular data protection and personal rights, especially in the EU. Furthermore, they point to a change in work, in which workers will move to having controlling or monitoring roles, while the value-creating work will be taken over entirely by machines. Thus, they think digital twins of workers would only have benefits for training or even none at all. From the point of view of some experts, the creation of a digital twin of a work person is too complex and only makes sense for machines. Others argue that the costs of creating a digital twin are very high, which would make their economic viability questionable for most companies. For some experts, however, the idea is simply too visionary. On the other hand, there are experts who estimate the probability of occurrence as high. Some consider the digital twin to be reasonable and feasible. Most of the experts, however, are more cautious and estimate that their introduction will take a long time and will be gradual: the reasons given for this are again the resistance of employees and unions, as well as the complexity of digital twins themselves. Changes in the way individuals and society deal with (their own) data are seen as the main drivers of the development toward digital twins.

Regarding firm impact, the estimates by the experts from Germany and the ROW are not far apart, and overall the experts consider this projection to have a medium impact ($M = 3.13$). Most of the experts who expect the firm impact to be low expect restrictions with regard to the target groups: some see a meaningful use of digital twins only among office workers and others only among workers of a certain age or experience. On the other hand, those experts who predict a strong firm impact point out that information on skills and competences is important for planning. They see a general benefit with regard to production.

Before the conclusion, the authors would like to discuss two expert statements. Firstly, there is a need for discussion regarding the nature of the digital twin: in it, the human being is represented not only physically but also in terms of their cognitive load. If this aspect was not considered, we would be dealing with a mechanistic view of human beings which understands them as a kind of machine and in which psychological factors cannot be covered at all. Secondly, the assertion that humans will no longer be engaged in value-creating activities at some point in the future is a vision that is propagated by some groups. But it is by no means the only conceivable vision of the future. There is disagreement and a variety of positions here.

The authors conclude that, on the one hand, implementing digital twins of humans has the potential to create adaptive workspaces that adjust to the needs of employees and, thereby, benefit production productivity, safety, and employees' health. On the other hand, the concept is associated with substantial privacy concerns and opens up possibilities of misuse of the collected data resulting, e.g., in workplace discrimination "justified by objective data" or responsibility gaps due to unwarranted reliance on faulty, data-based assessments. Implementing digital shadows of employees in a firm will require the acceptance of the work council and taking the organizational culture into account.

7 Projection 12: Employees' Rights

The projection "employees' rights" deals with the impact of data collection on employees' rights. Experts were asked whether adequate anonymization procedures for the protection of employees' personal rights will have been introduced for firms that collect data on personal performance and work patterns in the form of digital twins of their employees. This was rated as having a high probability ($M = 57.07\%$, $SD = 26.01\%$). At the same time, there was a large amount of dissent among the experts ($IQR = 5.00$). While some experts stressed that such data protection practices would be a prerequisite for human digital twins (P11) and would be enforced by unions as well as workers' rights legislation, others raised doubts that even anonymization procedures would be sufficient to protect employees' privacy. The survey found a medium firm impact of $M = 2.93$.

In order to obtain acceptance for AI applications like digital shadows of employees and to ensure an ethical and legal implementation of such ideas, it will be necessary to guarantee the protection of employees' privacy and sovereignty over the individual-related data (Todolí-Signes, 2019). This can be achieved through the interaction of different approaches. On the one hand, research and development are required to develop (technical) procedures that prevent the disclosure of personal data and prevent any possibility of reconstructing the original identity of the employee based on the obtained data. On the other hand, processes that strictly limit the accessibility of the data must be developed and established to further prohibit misuse. This also requires thorough education of employees regarding their rights and the measures to protect their privacy so that they can make informed decisions about providing the company with access to their data. This would be done through coordination and active involvement with work councils.

Considering the European General Data Protection Regulation, it must be assumed that realizing privacy by design (Cavoukian, 2009) will be a legal requirement for using human digital twins. Moreover, the implementation of sufficient privacy measures will play a crucial role in determining employees' acceptance of human digital shadows. Employees' acceptance will, in turn, be an important factor in determining their reliance on the system. Insufficient acceptance will lead to misuse or disuse of the implemented systems, impeding any benefits they could provide. Therefore, a potential strategy for improving this acceptance among employees, besides merely teaching them about the processes at the beginning of data collection, would be an ongoing effort to make the involved collection processes as transparent as possible. If employees can always easily access information about the collected datasets, this might positively affect their acceptance of their digital twins, since it could strengthen their trust in the ongoing processes.

The impact of employees' acceptance and attitude toward use of their data also highlights the possibility that guaranteeing adequate handling of employee data may become a crucial competitive advantage across companies for recruiting new employees. This is because new employees might choose to work for a company that respects their privacy rights as it positively affects their emotion-led evaluation processes.

8 Projection 13: Workforce Reduction

Due to the recent and ongoing developments in automation, ever more work originally performed by human workforces may become automated, and hence, that workforce may become redundant (Autor, 2015; Acemoglu & Restrepo, 2017; Brynjolfsson & Mitchell, 2017). According to the industry association International Federation of Robotics (IFR), Germany has the fifth highest robot usage in industry worldwide (IFR, 2021). With this trend continuing and with further technological advancements in robotics likely to happen, automation will have fundamental implications on the future workplace, especially in production. Furthermore, a World Economic Forum (WEF) survey from 2020 suggests that half of production work could be done by robots as early as by the year 2025. This may amount to the displacement of some 85 million jobs worldwide (WEF, 2020).

In the Delphi study, experts were asked whether they predict that AI-based software and robots will bring about a significant reduction in the workforce by the year 2030. In the survey, the experts from industry and academia agreed on a medium probability of a significant reduction, with an average expectation of 51.55% ($SD = 18.85\%$). However, they expect significant variability across the different industrial domains. Additionally, while our experts predicted that a significant reduction of the workforce would have a profound impact on enterprises and businesses ($M = 3.62$), there seems to be a consensus that new jobs will emerge and that the workforce will, at least partially, shift to other fields. In contrast to many other projections from this study, the experts agreed overwhelmingly on this question ($IQR = 2.00$).

The predicted reduction of the workforce is likely to cause large-scale job displacement and possibly unemployment, especially for low-qualified workers. Under the assumption that jobs requiring higher educational qualifications are less likely to be redundant in the near future, which is supported by the WEF's survey and by our experts' predictions, increasing automation and reliance on AI will place increasing pressure on companies to re- and upskill their workforce and on societies both to produce a qualified workforce and to deal with the social repercussions of this development. Educational goals of re- and upskilling workers create conflicts on generational levels. Consider the example of stock-keeping clerks who have been working at their workplace for the past four decades and are not far from retirement: many of them may be unwilling to change paths this late in their working career or would experience mental stress if forced to do so. This not only threatens their autonomy, but it is questionable whether they will indeed reach the levels of skill necessary to re-enter the job market as, e.g., machine learning specialists.

Thus, re- and upskilling may not suffice as approaches to dealing with job displacement. Furthermore, it is not clear that there will be sufficient demand for work to guarantee employment.

Hence, there is an urgent need to determine an ethical framework for dealing with displacement-induced unemployment to avoid distributive injustice caused by the loss of low-qualified jobs, as well as to respect the autonomy to choose one's preferred education and life path. A discussion on the meaning and the normative value of work is necessary to examine the non-monetary social costs of unemployment, as there are concerns that the loss of work would be accompanied by a loss of potential non-monetary benefits of work. On the one hand, some argue, work may produce valuable non-monetary goods, e.g., the development of skills and the production of valuable goods, as well as social contribution and recognition (Gheaus & Herzog, 2016) or community building (Estlund, 2003). Hence, approaches like universal basic income (UBI) would be insufficient solutions for displacement-induced unemployment as they do not replace the non-monetary value of work. However, UBI does not prevent one from performing charitable work. On the other hand, some argue that work is merely required for living, going hand in hand with distributive injustice, domination, unhappiness, and dissatisfaction. The absence of a socioeconomic requirement to work would thus enhance people's autonomy and even, some suggest, create "utopian possibilities and enable heightened forms of human flourishing" (Danaher, 2019). From this perspective, AI and robots might liberate workers from dangerous, unpleasant, or routine work, allowing them to work on more meaningful tasks voluntarily.

Notably, however, as the workforce was not predicted to be fully replaced by robots, further research is needed to explore how to design workplaces to enable productive and human-centered human-technology collaboration. Thus, the implications of workforce reduction should be further elaborated in light of the projections on autonomous robots, hybrid intelligence, AI assistants, and their implications.

9   Summary

In this chapter, we have outlined how the digital transformation of production, Industry 4.0, and our vision of Next Generation Manufacturing will influence how we work, how the organization of work within companies will change, and what implications this will have on employment and labor rights.

Projection 7 looked at the merging of workspaces of and closer collaboration between employees and production robots. Collaborative robots moving autonomously on the shop floor and interacting with human workers will most probably not dominate production scenarios by 2030. High investment costs and the lack of methods to easily program robots are seen as major hurdles. Thus, to transform conventional robotic production to autonomous production, robots, technologies, and AI methods must be (further) developed, programming barriers lowered, and costs reduced.

Projection 8 addresses the use of hybrid intelligence—the combination of human and artificial intelligence—for decision-making. These hybrid intelligences can be applied in production from the shop floor to strategic management decision-making. Experts predict this trend will both have a strong impact on the management profession in the future and be very likely to occur. Hybrid intelligence constitutes a form of human-machine interaction, raising questions for both academia and practitioners on the reliability and fairness of such systems and, consequently, on the attribution of responsibility in the context of shared tasks.

Projection 9 builds on the consideration of hybrid intelligence and explores the role of AI-based assistants in manufacturing. AI assistance has the potential to disruptively change current decision-making processes. While all the experts agree on the promising potential of AI assistance, the majority still expect humans to be the central decision-making authority who must consider ethical and legal implications. In consequence, there is disagreement regarding whether all decisions will be made by intelligent algorithms in 2030, with many experts believing this will only be the case for supporting processes.

Projection 10 examines the extent to which digital transformation will influence leadership roles in companies. Even though the experts rated the potential impact of changes in management and leadership processes as low, it is noticeable that artificial intelligence has already led to changes in management and leadership decision-making. This is especially noticeable when it comes to leadership decisions within HR. However, this also brings negative effects because most current AI systems tend to exclude social aspects from their processes of decision-making. Therefore, if AI systems are to be implemented in an ethically justifiable way that also incorporates social aspects, organizations that want to use such AI systems within their processes of management and leadership decision-making must make efforts to guarantee that those social aspects are incorporated into their AI systems.

Projection 11 focuses on the digital shadow as one of the central concepts of Next Generation Manufacturing and examines the opportunities and risks associated with creating digital shadows of employees and their behaviors. These digital shadows of employees can be considered a double-edged sword: on the one hand, the processing of data collected on workers allows the optimization of the workplace, which is intended to lead to optimal working conditions (personal advantages) and increased productivity (economic advantages); on the other hand, the ethical question arises as to whether the collection of the required data interferes too much with the workers' privacy (personal disadvantage).

Projection 12 addresses the question of what impact the digital transformation of production will have on working conditions and employee rights. While the experts agree that processes of anonymization are needed if firms or organizations collect data, they disagreed on whether exactly this anonymization can be guaranteed sufficiently. Furthermore, it is questionable whether employees will accept such data collection. Therefore, firms and organizations that want to collect data must invest in training and education processes and also design the whole collection process transparently. Additionally, it must be guaranteed that employees' have sovereignty over their individual-related data and that the firm or organization will not misuse the collected data.

Finally, projection 13 addressed the implications of automation on employment. With the trend of automation continuing and further technological advancement in robotics likely, fundamental implications for future workplaces, especially in production, need to be considered. Experts predict that a significant reduction of the workforce will have a profound impact on enterprises and businesses. There seems to be a consensus that new jobs will emerge and that the workforce will, at least partially, shift to other fields. This development needs to be considered from ethical, social, economic, and psychological angles to allow value-capturing workplaces to be designed.

In summary, this chapter investigated how Next Generation Manufacturing will change work and the organization of work inside and outside companies from experts' perspectives. Many of the changes are associated with significant benefits for the companies and for value creation, but many ethical, social, and legal issues are still insufficiently addressed and remain subjects for open research and innovation.

We have seen that collaborative robots and hybrid intelligence are impactful topics for the future, but also that implementation barriers, algorithmic biases, and questions of responsibility are still insufficiently addressed. Consequently, we must explore how humane yet productive future workplaces with collaborative robots and human-AI teams can be designed, what specific barriers to acceptance may arise, and how these can be mitigated, for example, by conducting co-design workshops. The problem of AI and algorithmic bias needs to be addressed by developing approaches to reduce algorithmic bias in the first place, by increasing the explainability and transparency of AI models and decision support systems (Barredo Arrieta et al., 2020), and, lastly, by increasing AI literacy and raising awareness for this issue and its consequences among developers, implementers, and end users (Long & Magerko, 2020). Although a new, AI-informed leadership culture is considered unlikely by the experts, corresponding approaches are already becoming reality. Therefore, there is an urgent need to address social aspects and fairness in this area and to ensure that today's companies with more traditional management concepts do not risk their viability. The experts further confirmed the benefits of collecting more data about employees and making it usable, which raises unresolved questions of privacy and data sovereignty. One approach to resolving this dilemma could be that companies develop, together with their employees, concepts for the use of personal data and to identify optimal privacy-utility tradeoffs that are harmonized with all stakeholders involved.

Undoubtedly, value creation will increase through automation, autonomous and collaborative robots, and AI assistants (Kagermann, 2015). Since this may come at the price of significant job losses, we need to be prepared by carefully exploring concepts such as universal basic income and their implications, as well as by continuously re- and upskilling employees in vulnerable positions, giving them the opportunities to acquire new competences and develop new skills (Brynjolfsson & Mitchell, 2017). However, this must not be imposed by the management, but developed together with employees and in line with their interests and capabilities to increase motivation, self-determination, and autonomy (Deci & Ryan, 2008). A central prerequisite for this is that we understand precisely the competences and skills needed for digitalized production and how these can be conveyed to prepare current employees for the future, ideally allowing employers and employees to engage in meaningful work. The digital transformation of production is inevitable. It remains a central challenge to shape this transformation with and for the people.

Acknowledgment Funded by the Deutsche Forschungsgemeinschaft (DFG, German Research Foundation) under Germany's Excellence Strategy – EXC-2023 Internet of Production – 390621612.

References

Autor, D. H. (2015). Why are there still so many jobs? The history and future of workplace automation. *Journal of Economic Perspectives, 29*(3), 3–30. https://doi.org/gc3cft.

Acemoglu, D., & Restrepo, P. (2017). The race between machine and man. *American Economic Review, 108*(6), 1488–1542. https://doi.org/hg54.

Barredo Arrieta, A., Díaz-Rodríguez, N., Del Ser, J., Bennetot, A., Tabik, S., Barbado, A., Garcia, S., Gil-Lopez, S., Molina, D., Benjamins, R., Chatila, R., & Herrera, F. (2020). Explainable artificial intelligence (XAI): Concepts, taxonomies, opportunities and challenges toward responsible AI. *Information Fusion, 58*, 82–115. https://doi.org/ggqs5w.

Binns, R. (2018). Fairness in machine learning: Lessons from political philosophy. In *Conference on fairness, accountability and transparency* (pp. 149–159). PMLR. https://doi.org/10.48550/arXiv.1712.03586

Borenstein, J. (2011). Robots and the changing workforce. *AI and Society, 26*, 87–93. https://doi.org/bkjdtk.

Brauner, P., Philipsen, R., Calero Valdez, A., Ziefle, M., & Philipsen, R. (2019). What happens when decision support systems fail?—The importance of usability on performance in erroneous systems. *Behaviour & Information Technology, 38*(12), 1225–1242. https://doi.org/ggm4rx.

Brauner, P., Dalibor, M., Jarke, M., Kunze, I., Koren, I., Lakemeyer, G., Liebenberg, M., Michael, J., Pennekamp, J., Quix, C., Rumpe, B., Van Der Aalst, W., Wehrle, K., Wortmann, A., & Ziefle, M. (2022). A computer science perspective on digital transformation in production. *ACM Transactions on Internet of Things (TIOT), 3*(2), 1–32. https://doi.org/hg56.

Brynjolfsson, E., & Mitchell, T. (2017). What can machine learning do? Workforce implications. *Science, 358*(6370), 1530–1534. https://doi.org/gcsjbw.

Brillowski, F., Dammers, H., Koch, H., Müller, K., Reinsch, L., & Greb, C. (2021). Know-how transfer and production support systems to cultivate the internet of production within the textile industry. In *International conference on intelligent human systems integration* (pp. 309–315). Springer. https://doi.org/hg47.

Brillowski, F., Greb, C., & Gries, T. (2020). Increasing the sustainability of composite manufacturing processes by using algorithm-based optimisation and evaluation for process chain design. *International Journal of Sustainable Manufacturing, 4*(2–4), 350–364. https://doi.org/hg57.

Castro, A., Silva, F., & Santos, V. (2021). Trends of human-robot collaboration in industry contexts: Handover, learning, and metrics. *Sensors, 21*(12), 4113. https://doi.org/hg58.

Cavoukian, A. (2009). Privacy by design the 7 foundational principles. https://www.ipc.on.ca/wp-content/uploads/resources/7foundationalprinciples.pdf.

Cummings, M. L. (2017). Automation bias in intelligent time critical decision support systems. In *Decision making in aviation* (pp. 289–294). Routledge. https://doi.org/10.2514/6.2004-6313

Danaher, J. (2019). *Automation and utopia*. Harvard University Press.

Deci, E. L., & Ryan, R. M. (2008). Self-determination theory: A macrotheory of human motivation, development, and health. *Canadian Psychology, 49*(3), 182. https://doi.org/fcnwzz.

Duggan, J., Sherman, U., Carbery, R., & McDonnell, A. (2020). Algorithmic management and app-work in the gig economy: A research agenda for employment relations and HRM. *Human Resource Management Journal, 30*(1), 114–132. https://doi.org/ggx8g8.

Engels, F. (1971). Die Lage der arbeitenden Klasse in England. Otto Wigand. Online. Accessed May 31, 2022, from https://books.google.at/books?id=B3BHAAAAYAAJ

Estlund, C. (2003). *Working together: How workplace bonds strengthen a diverse democracy*. Oxford University Press.

Gawer, A. (2014). Bridging differing perspectives on technological platforms: Toward an integrative framework. *Research Policy, 43*(7), 1239–1249. https://doi.org/gc8sc5.

Gheaus, A., & Herzog, L. (2016). The goods of work (other than money!). *Journal of Social Philosophy, 47*(1), 70–89. https://doi.org/f3m7sk.

Grau, A., Indri, M., Bello, L. L., & Sauter, T. (2017). Industrial robotics in factory automation: From the early stage to the Internet of Things. In IECON 2017-43rd Annual Conference of the IEEE Industrial Electronics Society (pp. 6159–6164). https://doi.org/gnkwrd

Grieves, M. (2011). *Virtually perfect: Driving innovative and lean products through product lifecycle management.* Space Coast Press. ISBN 978-0982138007.

Grieves, M., & Vickers, J. (2017). Digital twin: Mitigating unpredictable, undesirable emergent behavior in complex systems. In *Transdisciplinary perspectives on complex systems* (pp. 85–113). Springer. https://doi.org/gjnr49.

Gualtieri, L. (2021). *Methodologies and guidelines for the design of safe and ergonomic collaborative robotic assembly systems in industrial settings.* Dissertation Free University in Bolzano.

Hacker, P. (2018). Teaching fairness to artificial intelligence: Existing and novel strategies against algorithmic discrimination under EU law. *Common Market Law Review, 55*(4), 1143–1186. Available at SSRN https://ssrn.com/abstract=3164973

IFR. (2021). *Executive Summary World Robotics 2021.* International Federation of Robotics. Retrieved November 16, 2021, from https://ifr.org/freedownloads/

Institut Montaigne. (2020). Algorithms: Please Mind the Bias! Report – March 2020. Retrieved December 2, 2021, from www.institutmontaigne.org, https://www.institutmontaigne.org/en/publications/algorithms-please-mind-bias

Jarrahi, M. H., Newlands, G., Lee, M. K., Wolf, C. T., Kinder, E., & Sutherland, W. (2021). Algorithmic management in a work context. *Big Data & Society, 8*(2), 1–14. https://doi.org/gk3vkn.

Jones, D., Snider, C., Nassehi, A., Yon, J., & Hicks, B. (2020). Characterising the digital twin: A systematic literature review. *CIRP Journal of Manufacturing Science and Technology, 29,* 36–52. https://doi.org/ghg846.

Kagermann, H. (2015). Change through digitization—Value creation in the age of industry 4.0. In *Management of permanent change* (pp. 23–45). Springer Gabler. https://doi.org/gmcvf5.

Kellogg, K. C., Valentine, M. A., & Christin, A. (2020). Algorithms at work: The new contested terrain of control. *Academy of Management Annals 2020, 14*(1), 366–410. https://doi.org/ggdfs2.

Lerner, J. S., Le, Y., Valdesolo, P., & Kassam, K. S. (2015). Emotion and decision making. *Annual Review of Psychology, 66,* 799–823. https://doi.org/gdh74s.

Lepore, J. (2019). *Are robots competing for your job?* The New Yorker, Annals of Technology. Retrieved December 2, 2021.

Liao, Y., Deschamps, F., Loures, E. D. F. R., & Ramos, L. F. P. (2017). Past, present and future of industry 4.0-a systematic literature review and research agenda proposal. *International Journal of Production Research, 55*(12), 3609–3629. https://doi.org/ggwrtk.

Liehner, G. L., Brauner, P., Schaar, A. K., & Ziefle, M. (2021). Delegation of moral tasks to automated agents the impact of risk and context on trusting a machine to perform a task. *IEEE Transactions on Technology and Society,* 1–14. https://doi.org/hg6f.

Long, D., & Magerko, B. (2020). What is AI literacy? Competencies and design considerations. In *Proceedings of the 2020 CHI conference on human factors in computing systems* (pp. 1–16). https://doi.org/10.1145/3313831.3376727

Malik, A. A., Masood, T., & Kousar, R. (2021). Reconfiguring and ramping-up ventilator production in the face of COVID-19: Can robots help? *Journal of Manufacturing Systems, 60*(S), 864–875. https://doi.org/ghbz2q.

Mertens, A., Pütz, S., Brauner, P., Brillowski, F., Buczak, N., Dammers, H., ... & Nitsch, V. (2021). Human digital shadow: Data-based modeling of users and usage in the internet of production. In 2021 14th International Conference on Human System Interaction (HSI) (pp. 1–8). https://doi.org/hg6g.

Neff, G., & Nagy, P. (2016). Automation, algorithms, and politics| talking to bots: Symbiotic agency and the case of Tay. *International Journal of Communication, 10,* 4915–4931.

Neuburger, R., & Fiedler, M. (2020). Zukunft der Arbeit—Implikationen und Herausforderungen durch autonome Informationssysteme. *Schmalenbach Journal of Business Research, 9*(72), 343–369. https://doi.org/hg6h.

Ochmann, J., & Laumer, S. (2020). AI recruitment: Explaining job seekers' acceptance of automation in human resource management. In Proceedings of the 15th Proceedings of the 15th International Conference on Wirtschaftsinformatik. Potsdam. https://doi.org/hg6j.

Parker, S. K., & Grote, G. (2020). Automation, algorithms, and beyond: Why work design matters more than ever in a digital world. Applied Psychology. https://doi.org/ggjwzv.

Patel, V. L., Kaufman, D. R., & Arocha, J. F. (2002). Emerging paradigms of cognition in medical decision-making. *Journal of Biomedical Informatics, 35*(1), 52–75. https://doi.org/bqzbb2.

Parasuraman, R., & Manzey, D. H. (2010). Complacency and bias in human use of automation: An attentional integration. *Human Factors, 52*(3), 381–410. https://doi.org/fcmx4h.

Rani, U., & Furrer, M. (2021). Digital labour platforms and new forms of flexible work in developing countries: Algorithmic management of work and workers. *Competition and Change, 25*(2), 212–236. https://doi.org/ggmxmn.

Schemmer, T., Brauner, P., Schaar, A. K., Ziefle, M., & Brillowski, F. (2020), User-centred design of a process-recommender system for fibre-reinforced polymer production. In International conference on human-computer interaction (pp. 111–127). Springer. https://doi.org/hg6n.

Schuh, G., Prote, J. P., Gützlaff, A., Thomas, K., Sauermann, F., & Rodemann, N. (2019). Internet of production: Rethinking production management. In Production at the leading edge of technology (pp. 533–542). Springer Vieweg. https://doi.org/hg6p.

Skitka, L. J., Mosier, K. L., & Burdick, M. (1999). Does automation bias decision-making? *International Journal of Human-Computer Studies, 51*(5), 991–1006. https://doi.org/bg5rb7.

Subrahmanyam, A. (2013). Algorithmic trading, the flash crash, and coordinated circuit breakers. *Borsa Istanbul Review, 13*(3), 4–9. https://doi.org/hg6r.

Tambe, P., Cappelli, P., & Yakubovich, V. (2019). Artificial intelligence in human resources management: Challenges and a path forward. *California Management Review, 61*(4), 15–42. https://doi.org/gf6jxt.

Todolí-Signes, A. (2019). Algorithms, artificial intelligence and automated decisions concerning workers and the risks of discrimination: The necessary collective governance of data protection. *Transfer: European Review of Labour and Research, 25*(4), 465–481. https://doi.org/ggr6r5.

Villani, V., Pini, F., Leali, F., & Secchi, C. (2018). Survey on human–robot collaboration in industrial settings: Safety, intuitive interfaces and applications. *Mechatronics, 55*, 248–266. https://doi.org/cw6q.

Wachter, S., Mittelstadt, B., & Floridi, L. (2017). Why a right to explanation of automated decision-making does not exist in the general data protection regulation. *International Data Privacy Law, 7*(2), 76–99. https://doi.org/10.2139/ssrn.2903469

WEF. (2020). *The future of jobs report 2020.* World Economic Forum. Retrieved December 2, 2021, from http://www3.weforum.org/docs/WEF_Future_of_Jobs_2020.pdf

WGP. (2019). WGP-Standpunkt: KI in der Produktion—Künstliche Intelligenz erschließen für Unternehmen [WGP Viewpoint: AI in production—opening up artificial intelligence for companies] (2019) https://wgp.de/wp-content/uploads/WGP-Standpunkt_KI-final_20190906–2.pdf (last access 2022-02-18).

Zerilli, J., Knott, A., Maclaurin, J., & Gavaghan, C. (2019). Algorithmic decision-making and the control problem. *Minds and Machines, 29*(4), 555–578. https://doi.org/hg6t.

Capability Configuration in Next Generation Manufacturing

Christian Hinke, Luisa Vervier, Philipp Brauner, Sebastian Schneider, Linda Steuer-Dankert, Martina Ziefle, and Carmen Leicht-Scholten

Abstract Industrial production systems are facing radical change in multiple dimensions. This change is caused by technological developments and the digital transformation of production, as well as the call for political and social change to facilitate a transformation toward sustainability. These changes affect both the capabilities of production systems and companies and the design of higher education and educational programs. Given the high uncertainty in the likelihood of occurrence and the technical, economic, and societal impacts of these concepts, we conducted a technology foresight study, in the form of a real-time Delphi analysis, to derive reliable future scenarios featuring the next generation of manufacturing systems. This chapter presents the capabilities dimension and describes each projection in detail, offering current case study examples and discussing related research, as well as implications for policy makers and firms. Specifically, we discuss the benefits of capturing expert knowledge and making it accessible to newcomers, especially in highly specialized industries. The experts argue that in order to cope with the challenges and circumstances of today's world, students must already during their education at university learn how to work with AI and other technologies. This means that study programs must change and that universities must adapt their structural aspects to meet the needs of the students.

[Abstract generated by machine intelligence with GPT-3. No human intelligence applied.]

C. Hinke (✉)
Chair for Laser Technology, RWTH Aachen University, Aachen, Germany
e-mail: christian.hinke@llt.rwth-aachen.de

L. Vervier · P. Brauner · M. Ziefle
Human-Computer Interaction Center, RWTH Aachen University, Aachen, Germany
e-mail: vervier@comm.rwth-aachen.de; brauner@comm.rwth-aachen.de;
ziefle@comm.rwth-aachen.de

S. Schneider · L. Steuer-Dankert · C. Leicht-Scholten
Research Group Gender and Diversity in Engineering, RWTH Aachen University, Aachen, Germany
e-mail: sebastian.schneider@gdi.rwth-aachen.de; linda.steuer@gdi.rwth-aachen.de;
carmen.leicht@gdi.rwth-aachen.de

F. T. Piller et al. (eds.), *Forecasting Next Generation Manufacturing*, Contributions to Management Science, https://doi.org/10.1007/978-3-031-07734-0_6

1   Introduction

Industrial production systems are facing radical change in multiple dimensions. This change is caused by technological developments and the digital transformation of production, as well as the call for political and social change to facilitate a transformation toward sustainability. These changes affect both the capabilities of production systems and companies and the design of higher education and educational programs.

Using a novel real-time Delphi approach (see chapter "Applying the Real-Time Delphi Method to Next Generation Manufacturing" for a presentation of the method and the sample, as well as chapter "Big Picture of Next Generation Manufacturing" for an overview of the results), we developed propositions for different scenarios within Next Generation Manufacturing in 2030. As suggested by Gawer (2014), we used an integrative framework for platforms, distinguishing four dimensions: governance (e.g., open forms of collaboration; see chapter "Governance Structures in Next Generation Manufacturing"), organization (e.g., boundaries and decision-making; see chapter "Organization Routines in Next Generation Manufacturing"), capabilities (e.g., hybrid intelligence, this chapter), and interfaces (e.g., open APIs and human-machine interfaces; see chapter "Interface Design in Next Generation Manufacturing"). In addition, and influenced by our shared experiences during the COVID-19 pandemic, we added a fifth cluster of propositions, addressing the need for resilience in future digital manufacturing (see chapter "Resilience Drivers in Next Generation Manufacturing"). We provide a set of 24 validated projections based on 1930 quantitative estimations and 629 qualitative arguments from 35 industrial and academic experts from Europe, North America, and Asia. In so doing, we deliver a basis on which to substantiate academic discussions and which can support firm decision-making on future technological developments and economic implications that go beyond current speculations and siloed research.

In this chapter, we examine the effects that experts predict the transformation of production will have in relation to the capabilities of production systems (see Fig. 1). One core capability of a production system is the utilization of human work and especially of expert knowledge. We are currently seeing the digital transformation of expert knowledge into intelligent systems. This digital representation of explicit and implicit expert knowledge and its systematic transfer and processing could become a critical factor for success in the future of production. Projection 14 (expert knowledge) addresses these questions and highlights the implications for experienced workers.

These questions are closely linked to projection 17 (university degrees), which addresses the implications of these changes regarding academic education. In particular, projection 17 deals with the potential need for changes toward more holistic and interdisciplinary study programs in universities. Higher education institutions are already in the process of creating and introducing innovative multi- or interdisciplinary study programs. This change is characterized by a need for more interdisciplinary competencies and an extension of the engineering subjects to include

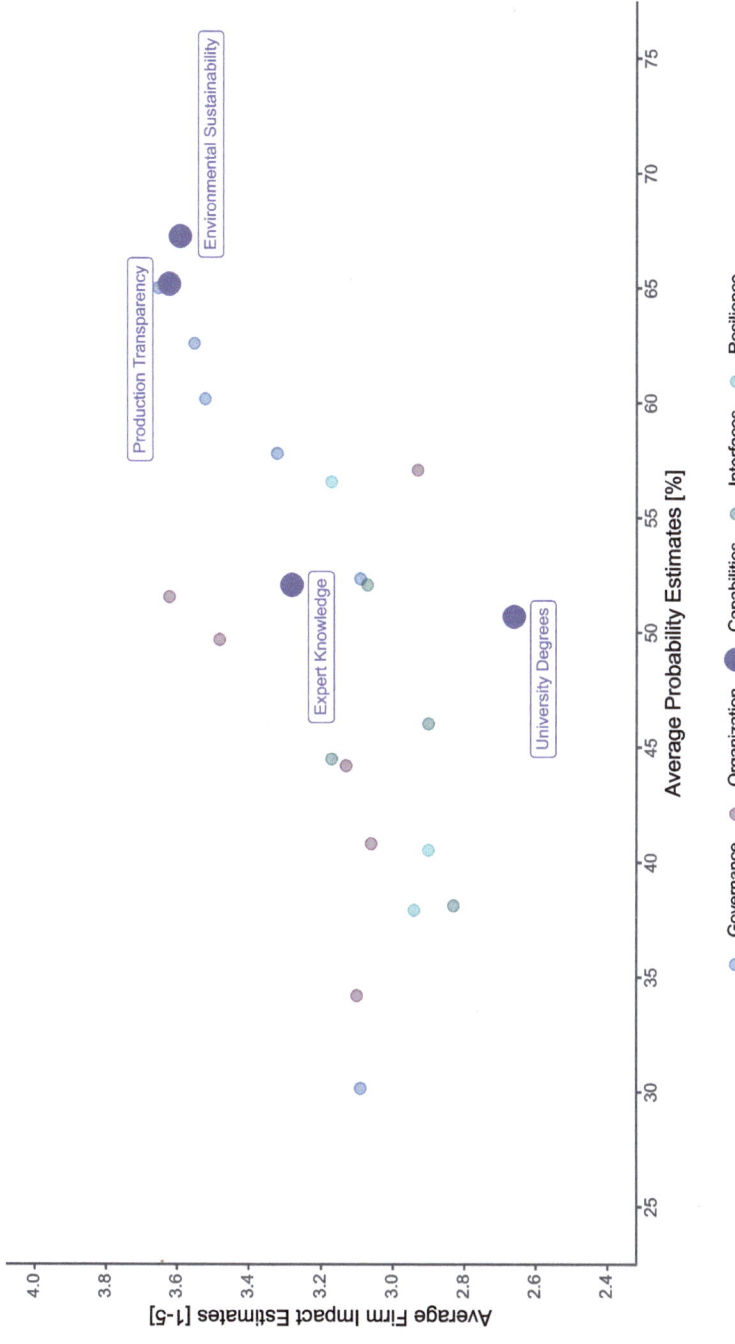

Fig. 1 Expert assessment of capability configuration in Next Generation Manufacturing (see chapter "Big Picture of Next Generation Manufacturing" for the full results of the Delphi survey)

computer science competencies. Projection 17 (university degrees) questions the demand for such developments.

Another core capability of a production system is its ability to produce the right products (effectivity) in the right way (efficiency). Regarding these aspects, a production system is traditionally defined by the conflicting criteria of quality, cost, and time. This triangle is increasingly being supplemented by sustainability as a fourth conflicting criteria. If sustainability became an equivalent or even the dominant criteria, this would change industrial production systems in a fundamental way. Projection 15 (environmental sustainability) questions whether environmental sustainability of production by 2030 will have increased significantly compared to today and illuminates the implications of this development.

The efficiency of a production system is closely related to the optimized allocation of all resources within the production system. Extensive and, as far as possible, full transparency is an essential prerequisite for this optimized allocation of resources. Projection 16 (production transparency) investigates whether full transparency based on a complete digital twin of all production machines, lines, and plant engineering and a complete digital shadow of their operations will increase production efficiency significantly.

2 Projection 14: Expert Knowledge

One vision regarding the advent of the Fourth Industrial Revolution is smart connections between intelligent systems as well as between them and humans. With the ability to learn and share data as well as to communicate with each other and collaborate autonomously, aspects such as performance, maintenance, manufacturing of individualized products, and generating transparency over the whole production process or value chain of a company will be improved (Zenkert et al., 2021).

One essential factor for a successful and well-functioning interaction is the digital transformation of expert knowledge into the systems. A widely agreed-upon definition of the term expert knowledge has not yet been established, but in general, it can be described as essential know-how on a specific issue that is not generally known by others. Expert knowledge can be classified into three types, namely, explicit, implicit, and tacit knowledge. Explicit knowledge can be expressed in words, numbers, and symbols and is easily communicated between individuals and organizations. Implicit knowledge refers to knowledge that is gained through incidental activities without awareness that learning is occurring. Tacit knowledge is gained from personal experience and is more difficult to express (Feng et al., 2017; Nonaka & Takeuchi, 2007).

Due to digital transformation in production, demographic change, and the shortage of skilled workers, capturing and making knowledge accessible is not new but has never before been so important and at the same time so challenging. This stems from the fact that expert knowledge, especially the implicit and tacit types, is challenging to capture. The other factor is that the differences in education and technological understanding between different generations have never been so great.

Regarding the ongoing industrial transformation, experts in the Delphi study were given a proposition that in 2030 implicit expert knowledge will increasingly be preserved explicitly in the form of digital models, interactive guides, or instructions and facilitated by technologies like augmented or virtual reality. Moreover, it was stated in the proposition that the knowledge would be also made available to novices and would eliminate the dependency on experienced production employees.

With an average expectation of 52.07% ($SD = 20.95\%$), the experts considered the proposition to be probable. The projected impact of digital knowledge capture on companies was $M = 3.28$. However, reasons given for the projection not occurring by 2030 were due to a lack of time, different understandings of expert knowledge, problems associated with generalizing algorithms, a lack of robust working technologies, and a lack of willingness to share expert knowledge.

The benefits of capturing knowledge and making it accessible to newcomers are that it adds a great deal of value, especially in highly specialized industries (e.g., textile industry). Outsourcing the learning of skills to digital technologies would be a great gain in efficiency and effectivity. Strengthening the robustness of such systems would bring significant benefits if knowledge was sustainably preserved and was retrievable by the company. However, a key question that needs to be solved is how systematic knowledge capture can be achieved. In one option, knowledge could be captured through complete digital documentation of decisions and process steps by employees. Another is to systematically extract expert knowledge through serious games and gamified environments, allowing more and also negative outcomes to be classified in a short time (Schemmer et al., 2022).

Another industrial example that underlines the importance of the digital transformation of expert knowledge can be found in laser-based production machines, which are highly complex systems. To be able to optimize the corresponding production processes or to detect expensive malfunctions and failures in advance, many years of experience and fundamental expert knowledge are necessary. Condition Based Services (CBS) by TRUMPF increase the availability and productivity of networked laser systems. CBS evaluates sensor data (> 250 sensors per machine) and detects risks and potential for improvement. By aggregating all the data in customer-specific dashboards, the system provides a solid knowledge and decision base, especially for non-expert machine users.

The systematic capture of explicit and implicit expert knowledge and its systematic transfer and processing could become critical factors for success in the future of production. Especially against the background of an aging workforce and an impending shortage of skilled workers, the need to deal with possibilities, benefits, barriers, and conditionals is increasing. In this context, an interdisciplinary approach is urgently required: in addition to labor law issues, it is necessary to consider how the accumulated knowledge can be applied in practice. This raises the question of how this knowledge can be translated into digital shadows, while concepts to bring this knowledge back into use should be developed. In this context, the question of suitable user interfaces is essential.

3   Projection 15: Environmental Sustainability

Projection 15 (environmental sustainability) deals with the concept of sustainable development, described in the Brundtland Report as "development that meets the needs of the present without compromising the ability of future generations to meet their own needs" (Brundtland, 1987). Environmental sustainability is currently playing a major role in the global political debate and is leading to major political initiatives, e.g., the Green New Deal in the USA and the European Green Deal in Europe (European Commission, 2019). Even if there is no generally accepted definition of environmental sustainability, most experts agree with the following two defining characteristics: (a) no net emissions of greenhouse gases and (b) economic growth decoupled from resource use. A major driver for the increased significance of environmental sustainability is the ongoing tightening of the legal framework in this field. In addition to these legal regulations, customers are rethinking and partly reducing their consumption behaviors in a society-wide trend. In addition, there are (financial) incentives to encourage the protection and preservation of common goods and to penalize environmental damage from products.

There is consensus that environmental sustainability of production will play a bigger role in 2030 ($IQR = 2.00$). The experts consider this projection to have a high probability ($M = 67.24\%$, $SD = 19.14\%$) and a high impact ($M = 3.59\%$). This is particularly remarkable as it has the highest probability of all projections. This high probability and high impact reflect the high significance of environmental sustainability in political and social discussions. It is interesting that the likelihood given by the industry experts ($M = 68.42\%$) is higher than that from academia ($M = 65.00\%$). This may reflect the objections from some of the academic experts that there is still more research on productivity than on sustainability.

Several experts stated that improvements in emerging and developing countries will be crucial for global environmental sustainability. Experts from the rest of the world expect a significantly higher average impact ($M = 4.00$) compared to experts from Germany ($M = 3.43$). In general, the experts see multiple drivers of this projection: climate goals of governments and regulation, requirements by investors and financial markets, and demand-driven selection of sustainable companies by customers.

In addition to the further tightening of limits and legal framework conditions, most experts interviewed in this Delphi study stated the importance of efficient use of resources, new technology, and innovative products for environmental sustainability of production. This demand is reflected by recent leading conferences like the Aachen Machine Tool Colloquium (Bergs et al., 2021) and by the new European research program for research and innovation (European Commission, 2021).

The implications of this projection on environmental sustainability of production can be summarized as follows: sustainable production and sustainable products (in terms of economic, ecological, and social sustainability) are demanded by the public and could provide a competitive advantage. Policy makers have started to encourage sustainability and establish societal transformation.

4 Projection 16: Production Transparency

Projection 16 (production transparency) deals with the concept of coupled digital twins and digital shadows along production lines. The increasing consideration of sustainability as an additional priority alongside the conflicting priorities of quality, cost, and time leads to a significant increase in the complexity of production systems. This projection states that full production transparency based on a complete digital twin of all production machines, lines, and plant engineering and a complete digital shadow of their operations should increase production efficiency significantly (Bergs et al., 2021).

The digital shadow and the digital twin are key enablers in this context. They connect all data and information linked to each asset of a production line across the full production line or system. According to the International Academy for Production Engineering (CIRP), a digital twin is a virtual image of a real device, object, machine, service, or immaterial process that describes its properties and behavior with the help of models, data, and information from within its life cycle (Stark & Damerau, 2019). A digital shadow, as defined by the Scientific Society for Production Engineering (WGP), is a sufficiently accurate representation of a production process with the purpose of creating a real-time capable evaluation basis of all relevant data (Bauernhansl et al., 2018).

Even though the experts in the Delphi study largely agree with this projection, giving it a quite high average probability ($M = 65.17\%$, $SD = 20.82\%$), there was no consensus among them that full production transparency could be achieved before 2030 ($IQR = 3.00$). Many experts claimed that 2030 is too early, and some others claimed that full production transparency could never be achieved for complex production systems. Experts consistently indicated that full transparency based on a complete digital twin of all production machines, lines, and plant engineering and a complete digital shadow of their operations will not be widely implemented by 2030. At the same time, most of the experts predicted a high firm impact of production transparency ($M = 3.62$). There is no significant difference between the industry experts and the academic experts regarding this assessment. However, there is a difference between the experts from Germany, who expect a significantly higher impact of this projection, and those from the rest of the world ($M = 3.76$ instead of $M = 3.25$). This may be related to the fact that many documented applications aiming for production transparency are implemented in Germany. The Condition Monitoring Center implemented by TRUMPF is a well-known example of production transparency. Based on the real-time analysis of networked sensor data (digital twin representing the laser system), the Condition Monitoring Center increases the technical availability and productivity of connected laser systems. Moreover, the digital twin can be opened up to external partners for collaboration. For example, Rolls-Royce built its R2 Data Labs as an ecosystem where it jointly innovates with external partners based on usage data from its engines.

Depending on the implementation, production transparency can simply provide production efficiency, but it can also be the basis for completely new business models which allow for further differentiation in the industry. Currently, only large firms have the computing power and data capabilities to make use of digital twins and production transparency, but this is likely to change. Decreasing costs for collecting, storing, and analyzing data will make the business case more favorable for many use cases.

5   Projection 17: University Degrees

Projection 17 (university degrees) deals with a potential need for changes toward more holistic and interdisciplinary study programs in universities. When asked whether, for example, the application of biological principles (e.g., cybernetics or biomimicry) in manufacturing will create a higher demand for multi- or interdisciplinary university degrees that incorporate the fields of engineering, life sciences, and computer science, experts disagreed ($IQR = 3.00$). Tellingly, such changes were rated as having a medium probability of occurring of 50.69% ($SD = 23.48\%$) and a medium level of firm impact of 2.66.

However, if we look at current changes in the study programs of universities, we can see that higher education institutions are already in the process of creating and introducing innovative multi- or interdisciplinary study programs. A qualitative study conducted by the GDI of RWTH Aachen University with the management of different companies from the production sector in 2019–2020 indicated a change in the requirements of job profiles. This change was characterized by a need for more interdisciplinary competencies and an extension of engineering subjects to include computer science competencies. Furthermore, diversity management competencies seem to be becoming increasingly important. Research has shown that, for example, within the context of engineering in a more and more globalized world, students must learn, already while they are studying, to work within teams that are marked by diversity and to become "globally competent engineers" (Downey et al., 2006; Leicht-Scholten et al., 2016; Leicht-Scholten & Steuer-Dankert, 2020; Dankert et al. 2019).

The importance of this is also apparent in connection with (P10) "new leadership" and the therein addressed need to pay attention to social aspects (e.g., diversity) surrounding the usage of AI decision software to avoid biased outcomes within HR decision-making processes. This can be achieved by offering students teaching formats that are characterized by interdisciplinary and intercultural teaching settings that provide space for critical reflections and discussions on diversity (Leicht-Scholten & Steuer-Dankert, 2020). This is, e.g., visible within the context of (P10) (see chapter "Organization Routines in Next Generation Manufacturing") because research introduces the fact that HR leaders are increasingly forced to understand processes surrounding the topics of AI and data science (Tambe et al., 2019) and, therefore, need to gain interdisciplinary knowledge and competencies (ideally while at university). As research further shows, such multi- and interdisciplinary approaches are also necessary to achieve more diversity-inclusive, fairer, and more usable (including from the perspective of other aspects of diversity like gender, race, and disabilities) technological outcomes that avoid discriminatory biases (Dankert et al., 2019; Leicht-Scholten & Steuer-Dankert, 2020). Therefore, processes of diversity, creativity, and skill management must also be integrated into study programs if, like in our example, humans and AI are to increasingly work as "teammates" (Dellermann et al., 2019 and see P9).

Students must learn already during their university education how they can cope with the challenges and circumstances of today's digitalized and globalized labor market. Since this means that study programs and their contents must change, it automatically also means that structural aspects of learning environments, or more generally universities themselves, must change. Yet this change is already visible within universities that are aware of the addressed needs for more holistic and interdisciplinary study programs.

6 Summary

In this chapter, we have examined the changing capabilities of industrial production systems driven by technological development (especially digital transformation) and political or social change (especially sustainability transformation). In the Delphi study, we investigated whether and in what ways capabilities related to human work and, especially, expert knowledge (P14), as well as the closely related academic education of expert workers (P17), will change in the future. We also investigated whether the traditional triangle of quality, cost, and time will be supplemented by sustainability as a fourth conflicting criteria (P15). Additionally, we outlined how full transparency-based digital twins and digital shadows will increase production efficiency (P16).

Projection 14 (expert knowledge) addresses the digital representation of explicit and implicit expert knowledge and outlines that the systematic transfer and processing of knowledge could become a critical but currently untapped factor for success in the future of production: an impending shortage of skilled workers requires firms to deal with possibilities, benefits, barriers, and conditionals of knowledge capture in the workplace. Industry and academia need to develop approaches that enable systematic knowledge capture, for example, by creating "knowledge-to-data" pipelines that capture fluid or tacit knowledge implicitly during work. This digitalized knowledge must then be made actionable as training material for the re- and upskilling of employees, for training AI-based decision support systems (Schemmer et al., 2020), or for developing automated control systems that are based on prior human behavior. This requires interdisciplinary approaches to develop systems that are compliant with labor laws, respect the privacy concerns of employees, and make the accumulated knowledge actionable. This raises the question of how this knowledge can be translated into digital traces and shadows and how these can be integrated within usable, actionable, and trusted support systems.

On projection 15 (environmental sustainability), we found a broad consensus of all experts that environmental sustainability of production will play a bigger role in 2030 and will have a high firm impact. This reflects the high significance of environmental sustainability in political and social discussions. In addition to the further tightening of limits and legal conditions, most experts interviewed in our Delphi study stated the importance of the efficient use of resources, new technologies, and innovative products for environmental sustainability of production. Several experts stated that improvements in emerging and developing countries will be crucial for global environmental sustainability. Some other experts from academia remarked that there is still more research on productivity than on sustainability. From a community point of view, sustainable production is demanded by the public and could provide a competitive advantage.

Projection 16 (production transparency) addresses full production transparency based on a complete digital twin of all production assets and a complete digital shadow of their operations. There is consensus among the experts that this full transparency could increase production efficiency significantly, but there is no consensus that full production transparency will be achieved by 2030. Besides simply improving production efficiency, full production transparency could also be the basis for completely new business models which allow for further differentiation in the industry.

How production transparency and traceability can be increased along supply chains while preserving the autonomy and privacy of companies is still an open challenge. While blockchains are suggested as possible approaches (Pennekamp et al., 2020), how these concepts can be implemented by both large enterprises and small- and medium-sized companies with smaller development budgets is still an open question.

Projection 17 (university degrees) states that, for example, the application of biological principles (e.g., cybernetics or biomimicry) in manufacturing will create a higher demand for multi- or interdisciplinary university degrees that incorporate the fields of engineering, life sciences, and computer science. There was disagreement among the experts ($IQR = 3.00$). Nevertheless, as actual research shows, such multi- and interdisciplinary approaches are necessary to achieve more diversity-inclusive, fairer, and more usable (including from the perspective of other aspects of diversity like gender, race, and disabilities) technological outcomes that avoid discriminatory biases (Dankert et al., 2019; Leicht-Scholten & Steuer-Dankert, 2020). Therefore, processes of diversity, creativity, and skill management must also be integrated into study programs.

The upcoming digital transformation of production will offer exciting new possibilities and will require new capabilities from companies, supply chains, and educational institutions. This chapter examined the experts' assessments of the impact and likelihood of occurrence of the technologies, tools, and methods developed by Next Generation Manufacturing on selected capabilities of production systems.

We showed that beyond evolutionary improvements in production technology, the capabilities of production systems will be shaped by three major trends: digital transformation of production, demographic change, and transformation toward sustainability. However, these challenges require holistic and inter- and transdisciplinary cooperation instead of siloed disciplinary approaches.

Acknowledgment Funded by the Deutsche Forschungsgemeinschaft (DFG, German Research Foundation) under Germany's Excellence Strategy – EXC-2023 Internet of Production – 390621612.

References

Bauernhansl, T., Hartleif, S., & Felix, T. (2018). The digital shadow of production—A concept for the effective and efficient information supply in dynamic industrial environments. *Procedia CIRP, 72*, 69–74.

Bergs, T., Brecher, C., Schmitt, R., & Schuh, G. (2021). Internet of production—Turning data into sustainability. In *Proceedings Aachener Werkzeugmaschinen-Kolloquium (AWK) 2021* (p. 466). Apprimus Verlag.

Brundtland, G. (1987). Our common future. Report of the World Commission on Environment and Development. United Nations.

Dankert, L. S., Gilmartin, S. K., Muller, C. B., Dungs, C., Sheppard, S. D., & Leicht-Scholten, C. (2019). Expanding engineering limits: A concept for socially responsible education of engineers. *The International Journal of Engineering Education, 35*(2), 658–673.

Dellermann, D., Ebel, P., Söllner, M., & Leimeister, J. M. (2019). Hybrid intelligence. *Business & Information Systems Engineering, 61*(5), 637–643. https://doi.org/10.1007/s12599-019-00595-2

Downey, G. L., Lucena, J. C., Moskal, B. M., Parkhurst, R., Bigley, T., Hays, C., . . . Nichols-Belo, A. (2006). The globally competent engineer: Working effectively with people who define problems differently. *Journal of Engineering Education, 95*(2), 107–122.

European Commission (2019). The European Green Deal. Communication from the commission to the European Parliament, the European Council, the council, the European Economic and Social Committee and the committee of the regions. https://www.eea.europa.eu/policy-documents/com-2019-640-final

European Commission (2021). Horizon Europe, the EU research and innovation programme (2021–27): For a green, healthy, digital and inclusive Europe. https://op.europa.eu/en/publication-detail/-/publication/93de16a0-821d-11eb-9ac9-01aa75ed71a1

Feng, S. C., Bernstein, W. Z., Hedberg, T., & Barnard Feeney, A. (2017). Toward knowledge management for smart manufacturing. *Journal of Computing and Information Science in Engineering, 17*(3), 031016.

Gawer, A. (2014). Bridging differing perspectives on technological platforms: Toward an integrative framework. *Research Policy, 43*(7), 1239–1249.

Leicht-Scholten, C., Steuer, L., & Bouffier, A. (2016). Facing future challenges: building engineers for tomorrow. In New Perspectives in Science Education Conference Proceedings (pp. 32–37).

Leicht-Scholten, C., & Steuer-Dankert, L. (2020). Educating engineers for socially responsible solutions through design thinking. In *Design thinking in higher education* (pp. 229–246). Springer.

Nonaka, I., & Takeuchi, H. (2007). The knowledge-creating company. *Harvard Business Review, 85*(7/8), 162.

Pennekamp, J., Alder, F., Matzutt, R., Mühlberg, J. T., Piessens, F., & Wehrle, K. (2020). Secure end-to-end sensing in supply chains. In 2020 IEEE Conference on Communications and Network Security (CNS) (pp. 1–6).

Schemmer, T., Brauner, P., Schaar, A. K., Ziefle, M., & Brillowski, F. (2020). User-centred design of a process-recommender system for fibre-reinforced polymer production. In *Human interface and the management of information, HCII 2020* (pp. 111–127). Springer. https://doi.org/10.1007/978-3-030-50017-7

Schemmer, T., Reinhard, J., Brauner, P., & Ziefle, M. (2022). Advantages and challenges of extracting process knowledge through serious games. In Proceedings GamiFIN Conference 2022. http://ceur-ws.org/Vol-3147/paper2.pdf

Stark, R., & Damerau, T. (2019). Digital twin. In S. Chatt & T. Tolio (Eds.), *CIRP encyclopedia of production engineering* (2nd ed.). Springer.

Tambe, P., Cappelli, P., & Yakubovich, V. (2019). Artificial intelligence in human resources management: Challenges and a path forward. *California Management Review, 61*(4), 15–42.

Zenkert, J., Weber, C., Dornhöfer, M., Abu-Rasheed, H., & Fathi, M. (2021). Knowledge integration in smart factories. *Encyclopedia, 1*(3), 792–811.

Interface Design in Next Generation Manufacturing

Ralph Baier, Srikanth Nouduri, Luisa Vervier, Philipp Brauner, István Koren, Martina Ziefle, and Verena Nitsch

Abstract With the advent of Next Generation Manufacturing, information and communications technologies have become an essential part of the production process, creating and providing data for all stakeholders. Given the high uncertainty in the likelihood of occurrence and the technical, economic, and societal impacts of associated transformations in production, we conducted a technology foresight study, in the form of a real-time Delphi analysis, to derive reliable future scenarios featuring the next generation of manufacturing systems. This chapter presents the interfaces dimension and describes each projection in detail, offering current case study examples and discussing related research, as well as implications for policy makers and firms. Interfaces play a major role in the provision of information. We discuss the trend of implicit user interfaces and the benefits of working from home. Implicit user interfaces are based on user inputs that are not directly aimed at giving a command, but are nevertheless captured, understood, and used by the computer system to provide a richer user experience. Working from home has many benefits, including reducing costs and dependencies. However, experts disagree on whether plant directors will manage multiple factories centrally via telework due to complete and real-time transparency of all operations in a digital system by 2030. The COVID-19 pandemic has shown that it is important to have such an infrastructure even if working from home may not be considered appropriate in many manufacturing companies. Mobile apps that support production management are one key issue in this context.

R. Baier (✉) · S. Nouduri · V. Nitsch
Institute of Industrial Engineering and Ergonomics, RWTH Aachen University, Aachen, Germany
e-mail: r.baier@iaw.rwth-aachen.de; s.nouduri@iaw.rwth-aachen.de; v.nitsch@iaw.rwth-aachen.de

L. Vervier · P. Brauner · M. Ziefle
Human-Computer Interaction Center, RWTH Aachen University, Aachen, Germany
e-mail: vervier@comm.rwth-aachen.de; brauner@comm.rwth-aachen.de; ziefle@comm.rwth-aachen.de

I. Koren
Chair of Process and Data Science, RWTH Aachen University, Aachen, Germany
e-mail: koren@pads.rwth-aachen.de

[Abstract generated by machine intelligence with GPT-3. No human intelligence applied.]

1   Introduction

With the advent of Next Generation Manufacturing, information and communications technologies have become an essential part of production processes, creating and providing data for all stakeholders (Brauner et al., 2022). Interfaces play a major role in the provision of information. Although interfaces have always existed in industry and production, they have changed due to the development of automation technologies (Papcun et al., 2018). Nowadays, there is a distinction between internal and external interfaces. Whereas internal interfaces are usually extensions of a production system controls expanded to include additional process and control functionality, external interfaces serve to connect the production system with the surrounding production facility. Internal interfaces are mainly systems for user interaction, such as human-machine interfaces (HMI), machine and data acquisition, or production management (Weck, 2006). There are various types of HMI, depending on the field of application and the degree of automation or, conversely, human influence (Gorecky et al., 2014). External interfaces are considered the connection between the production system and the environment. Against the background of these two perspectives, manufacturing firms face high uncertainties regarding the management and design of open external interfaces and internal HMI. Dealing with these uncertainties is the topic of this chapter (see Fig. 1).

Using a novel real-time Delphi approach (see chapter "Applying the Real-Time Delphi Method to Next Generation Manufacturing" for a presentation of the method and the sample, as well as chapter "Big Picture of Next Generation Manufacturing" for an overview of the results), we developed propositions for different scenarios within Next Generation Manufacturing in 2030. As suggested by Gawer (2014), we used an integrative framework for platforms, distinguishing four dimensions: governance (e.g., open forms of collaboration; see chapter "Governance Structures in Next Generation Manufacturing"), organization (e.g., boundaries and decision-making; see chapter "Organization Routines in Next Generation Manufacturing"), capabilities (e.g., hybrid intelligence; see chapter "Capability Configuration in Next Generation Manufacturing"), and interfaces (e.g., open APIs and human-machine interfaces; see this chapter). In addition, and influenced by our shared experiences during the COVID-19 pandemic, we added a fifth cluster of propositions addressing the need for resilience in future digital manufacturing (see chapter "Resilience Drivers in Next Generation Manufacturing"). We provide a set of 24 validated projections based on 1930 quantitative estimations and 629 qualitative arguments from 35 industrial and academic experts from Europe, North America, and Asia. In so doing, we deliver a basis on which to substantiate academic discussions and which can support firm decision-making on future technological developments and economic implications that go beyond current speculations and siloed research.

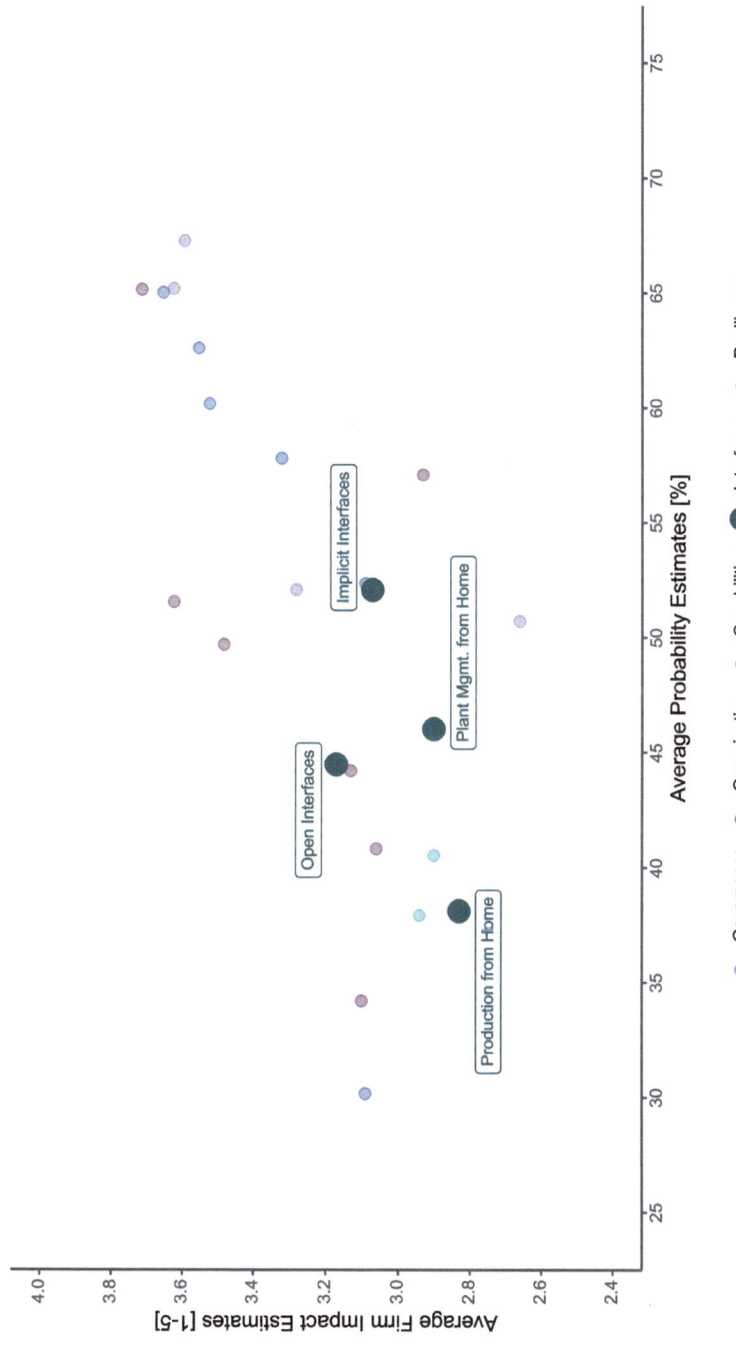

Fig. 1 Expert assessment on interface design in Next Generation Manufacturing (see chapter "Overview of Next Generation Manufacturing for the full results of the Delphi survey")

2   Projection 18: Implicit Interfaces

How we interact with computers and technology evolves constantly. While for decades the primary input devices have been keyboards and mouses (Dix et al., 2003), touch surfaces and voice assistants are now pervasive. In the near future, additional input and output devices, such as proximity or presence sensors and camera systems that can detect one's posture, intentions, fatigue, or mood, will enrich how we interact with technology (Brunner et al., 2021; Garcia-Garcia et al., 2017). Furthermore, while traditionally software systems have offered different users the same functionality and appearance, they have now become increasingly personalized. For example, past search queries, favorite songs, and shopping behaviors are integrated to give a better understanding of what we as users want and to provide search results, playlists, or shopping recommendations that better match our interests.

While these developments are mainly driven by consumer products and services, we assume they will continue and that there will be more implicit user interfaces in the future, including in production. Explicit user interfaces are ones where users interact with computers through direct commands – by either command line, speech, or a graphical user interface. In contrast, implicit interfaces are based on user inputs that are not intended to give a direct command, but are nevertheless captured, understood, and used by the computer system to provide a richer user experience (Schmidt, 2000).

Possible application areas of implicit user interfaces include human-robot collaboration (HRC) or decision support systems at the administrative level of manufacturing companies. For example, a robot at a hybrid workstation could detect the current workload of its human colleague utilizing a camera system and adjust its speed accordingly (Prewett et al., 2010). In the case of a decision support system for administrative tasks, the interface, the presentation of information, and the level of automation could be implicitly customized to best fit the human decision-maker's capabilities and current state and thereby reduce errors and workload (Kaber & Endsley, 2004).

AI-based approaches to both the recognition of human intent and selecting appropriate responses are being worked on to enable smooth and implicit interactions between people and technology. Despite much progress in the field of implicit interactions in production, many questions remain unresolved. For example, it is still a fundamental challenge for machines to reliably recognize human intent and to respond appropriately. Furthermore, transparency, traceability, security, reliability, trust, and social acceptance are unresolved issues.

Projection 18 states that in 2030, human-machine interaction will have evolved away from explicit interaction, where the human operator has full control of the actions of the production system's entities, toward implicit interaction, where the system automatically adapts to the human operator's behavior by detecting and predicting their actions and modifying these actions accordingly.

On average, the members of our expert panel assessed this trend as having a medium probability of occurrence by 2030 ($M = 52.07\%$, $SD = 17.93\%$) and a medium projected impact on companies ($M = 3.07$). However, there is disagreement among the experts on the likelihood of occurrence of implicit HMI by the year 2030. Consequently, there are both strong supporters and rejecters of this projection. According to the experts from our panel, the challenge lies in the complete development of implicit HMI, and 2030 is seen as most certainly too soon for this. Another problem is long machine lifecycles in manufacturing and thus slower adoption of new production machines that would explore the potential of future implicit interfaces. In addition, numerous unresolved issues regarding the social acceptance of implicit interfaces must be addressed by academia and industry. From the development of explicit interfaces for Next Generation Manufacturing, we already know that interface design, privacy perception, and trust in technology play essential roles in successful interaction and that human factors must be considered from the beginning of the technology's development (Hoff & Bashir, 2015; Valdeza et al., 2015; Brauner et al., 2022). Implicit interfaces for Next Generation Manufacturing must be designed to support employees in their work while being aligned with their expectations, norms, and values. Approaches to this include new implicit and explicit interfaces in Next Generation Manufacturing being developed in a stakeholder-orientated way together with employees (user-centered and participatory design) to identify and mitigate possible acceptance and interaction barriers at an early stage. Otherwise, companies risk losing the motivation of their employees through a lack of perceived utility, autonomy, and self-determination (Deci & Ryan, 2008), all of which are urgently needed for the upcoming digital transformation of production.

3 Projection 19: Open Interfaces

To enable data-driven services across organizational boundaries within Next Generation Manufacturing, the IT systems of two companies must communicate with each other to exchange data. Interfaces act as points of contact between involved stakeholders and are thus technically the foundation of distributed computing. Service-oriented architectures allow enterprises to encapsulate business functions in well-defined components. A look back at history reveals the role of open interfaces as a catalyst for the internet and computers in general. A computer with all its many different components from different manufacturers would not work without commonly defined and used interfaces. The internet makes this clear at the level of communication. It was only by means of standardized protocols that it became possible for all kinds of devices to communicate with each other via a local or worldwide network. Open interfaces are considered vital in Next Generation Manufacturing (cf. https://openindustry4.com/), as they make data exchange between manufacturers, customers, and service providers possible. The German

Standardization Roadmap Industrie 4.0 mentions that harmonized interfaces presuppose that these are based on standards and coordinated specifications.

Among the experts surveyed as part of the Delphi study, there is clear disagreement on the regulatory requirements for introducing open interfaces by the year 2030. The probability of this projection is estimated to be 44.48%, with a comparatively low standard deviation ($SD = 21.77\%$). The experts predicted that it would have a medium impact on firms ($M = 3.17$). The experts see two ways in which open interfaces can be established: through regulations or through (proprietary) standards from companies that become established on the market. The feasibility of implementation in the next 10 years is considered unrealistic, not least because resistance is expected from companies that want to protect their know-how. Nevertheless, the experts see open interfaces as a game-changer and a way to reduce costs and dependencies.

In the future, this certainly means that individual manufacturers will have to keep their products at least compatible with open interfaces to remain competitive. Companies will not risk running into a dead-end of incompatibility with their machines or robot fleets. An example of an open interface with standardized communication is the OPC Unified Architecture (OPC UA), a machine-to-machine communication protocol for industrial automation. It is freely available and implementable under the GPL 2.0 license and focuses on communication between industrial equipment and data collection and control systems.

Benefits of standardized interfaces in Next Generation Manufacturing include easier integration of products into production networks, coherent documentation, information about interaction possibilities, and streamlined modularity between components of various manufacturers. Some general concerns regarding interfaces include security-related aspects like possible entry gates for hackers and exploiters. Overly complex standards or overlooking configuration basics can also be potential problems. A prime example of this is keeping the default settings for access control (Dahlmanns et al., 2020).

Secure and sovereign data exchange among industrial partners is of overall strategic importance. The International Data Spaces Consortium is working on standardized processes, metamodels, and a technical reference architecture to enable data exchange. These go beyond pure (technical) descriptions of data exchange and enable new deployment scenarios, such as the inclusion of an independent third party to monitor and regulate policy-based data sharing between two organizations (see chapter "Governance Structures in Next Generation Manufacturing").

4   Projection 20: Production from Home

Rapid advancements in digital technologies for industrial applications are promoting the transformation of conventional production facilities into cyber-physical systems (CPS; Schumacher et al., 2020). Although these modern technologies have great application potential and have shown promising results in the research phase, their development as of today is not yet sufficient to meet real-world challenges. Therefore, it is farfetched to expect that all the operations at a production facility will be compatible with employees working from remote locations by 2030. The expert survey resulted in disagreement as to whether production employees will operate their workstations from home ($IQR = 3.00$), and the experts in this field do not expect that employees will predominantly work remotely by the year 2030 ($M = 38.10\%$, $SD = 24.93\%$). As well as this low probability rating, the experts do not believe that remote work will have a significant impact on the future of production ($M = 2.83$). While experts from the industry estimated this probability to be 37%, experts from academia estimated it to be 40%. Additionally, the experts from Germany gave lower estimations for this projection (probability: 33.10%; firm impact: 2.62) compared to those coming from other parts of the world (probability: 51.25%; firm impact: 3.38).

The COVID-19 pandemic has shown that there are situations for the industries where working from home can still help maintain production. Siemens, for example, has used this experience as the basis for its "New Normal Working Model" policy (Siemens, 2020). This agreement allows employees to work from home 2–3 days a week. Remote working can be beneficial for those who can utilize flexibility in their schedules to improve work productivity. While tasks, like modeling, simulations, documentation, etc., can be completed on a workstation from a remote location, the physical presence of a substantial portion of the workforce is required for shop floor operations. Although industries currently employ advanced robotic solutions for ongoing operations, remotely controlling artificial intelligence-based agents (cognitive robots, cobots) and having them perform comparably well to natural ones (human workers) still pose difficult challenges (see also (P7) to (P9) in chapter "Organization Routines in Next Generation Manufacturing").

For instance, in an automotive manufacturing facility, industrial-grade robots handle tedious tasks like material handling (transferring, stacking), processing operations (welding, painting, assembly), and final inspections. At the same time, a worker on the shop floor takes care of the tasks that require human intervention (system maintenance, machine tools setup, equipment repair). The ability of such experienced technicians to work remotely mainly depends on the need to use specialized equipment and make machine-assisted decisions. Closely related to the question of the systematic acquisition of expert knowledge (P14), it will be necessary to determine the nature of the information (visual or audiovisual) a worker needs to interpret so that this information can be incorporated into the decision-making process. Therefore, fundamental research that focuses on remote interactions between humans and production systems is needed to enable the execution of production tasks from workspaces at home in the future (Lund et al., 2020). A better understanding of human cognition is a primary requirement for passing these milestones, and this can be based on work in the fields of human-robot interaction, information visualization, and interface design.

As research in these fields is still embryonic, it is not fundamentally excluded that remote work will take place from beyond the shop floor, but it is only occasionally assumed that this can happen via working from home. At the same time, some see working from home as an option for tasks in more strategic areas, and industries are currently progressing in the direction of so-called hybrid offices, a mixture of telework and presence work. Looking at the reasons given for moving toward hybrid offices reveals a wide range of arguments, such as that human beings are social animals that rely on day-to-day interactions to survive and thrive. However, as part of the changes associated with Next Generation Manufacturing, Schwab (2017) points out that combining different technologies like robotics, mixed reality, and artificial intelligence could blur the lines between the physical and digital worlds, which would have a substantial influence on the workplaces of the future.

5   Projection 21: Plant Management from Home

In this projection, the perspective has changed from shop floor operations to that of the higher-level management. Here, the addressed question is whether, by 2030, plant directors will manage multiple factories centrally via telework due to complete and real-time transparency of all operations in a digital system. The experts from the Delphi study rated the probability of occurrence as medium ($M = 46.03\%$, $SD = 25.61\%$) and the firm impact as low ($M = 2.90$). Experts from both industry and academia gave similar estimations. While the German experts estimated the probability for this projection to be only 39%, experts from the rest of the world estimated it to be over 64%. Additionally, the firm impact was as low as 2.67 in Germany and higher in the rest of the world, at 3.50. There is strong disagreement among the experts on this topic ($IQR = 4.00$).

The differing opinions of the domain experts elicited several competing arguments. On the one hand, experts who estimate a lower probability of this projection emphasized two reasons: the timeline requirement for this implementation and the importance of human-to-human interactions among the top management. On the other hand, those experts who foresee a higher probability of occurrence stated that many aspects of this hypothesis already exist today and the implementation of remote plant management is more likely than remote production. Among the many infrastructural and technological hindrances mentioned, security risks associated with remote desktop connectivity and installing communication channels in residential buildings are prominent. Also, studies have long shown that frequent in-person interactions can lead to commitment, support, and cooperation among people on teams (Fayard et al., 2021). For some experts, this approach is incorrect as they believe that factory management is all about addressing problems directly on-site, and therefore handling such issues from home could be ineffective. Although it is also argued that the feasibility of remote management depends on the size of the factory, simpler tasks such as machine status queries can be handled remotely by the factory management, and facilitating such provisions would increase technological progress in this area.

The recent COVID-19 pandemic has shown that it makes sense to have such an infrastructure even if working from home may not be considered appropriate in many manufacturing companies. Mobile apps that support production management are one key issue in this context. Monitoring machine statuses in a digital twin of the factory via apps that support virtual execution of shop floor operations is a promising application prospect. Here, the key question that arises is which tasks can be carried out via telework, and it turns out, from the investigation of Lund et al. (2020), that it is not easy to answer this question. It is no surprise that working from home can affect highly qualified, well-educated employees and an increase or decrease in their productivity depends on the type of the task (Wu & Chen, 2020). However, Lund et al. (2020) also found that more than 20% of employees can work from remote locations for more than 2 days a week while being just as productive as if they were working from their offices.

Mobile applications for production management promise more flexibility and up-to-dateness, irrespective of the use case (telework or presence work). An optimized yet simplified decision-making process can be expected if users are presented with the right information at the right time and in the right way via a mobile application. Decision support systems in production can add real value to the process if they succeed in handling data from analysis using artificial intelligence-based agents to provide meaningful presentations for people in real time. Similarly, other aspects, like automated support, questions of liability, etc., will also become more relevant in this context.

6 Summary

In this chapter, insights about different types of human-machine interfaces were discussed, based on the results of the expert panel.

In projection 18, the question was raised of whether implicit interfaces will become established in the industry and whether they promise added value. Our expert panel could not provide a clear answer to the question. The typically long service lives of machines are seen as a particular obstacle, which noticeably slows down innovations in this area. This projection will depend on other factors, such as the retrofittability of the implicit interfaces. Retrofittability has worked before, for example, with numerical displays on production machines such as milling machines and lathes. Another aspect may be the attractiveness of the workplace with respect to the impending shortage of skilled workers. If implicit interfaces make work easier, then higher motivation, lower downtimes, and a lower error rate can also be expected. These inevitably lead to increased productivity, which is why retrofitting or even renewing the machines can pay off.

The question of whether, in 2030, regulatory requirements will demand open and standardized interfaces for data exchange for all kinds of manufacturing equipment was expressed in projection 19. There is disagreement among the experts regarding this projection, although they agree that open interfaces can only be introduced through legal regulations or if a manufacturer prevails on the market. However, it should not be underestimated what is happening in universities and other organizations in the field of open-source projects. A good example of this is the Robot Operating System (ROS), an open-source robotics middleware that originated at the Stanford Artificial Intelligence Laboratory as part of the Stanford AI Robot Project (STAIR) and is now widely used and making its way into the industry.

Projection 20 dealt with the question of whether, in 2030, production employees will operate their workstations from their homes, using remotely operated robots. The general question about working from home at the shop floor level cannot be answered conclusively: it is necessary to look in much more detail to identify which activities could be carried out remotely from home. For some of the interviewed experts, this will simply not be possible. For the rest of the experts, it is then necessary to clarify the extent to which teleworking makes work easier for the employee. Telework is not in itself a reason for enterprises to introduce it. It is rather a question of the workplace's attractiveness, flexible working hours, or prevention of infections or injuries. Many researchers believe there will be a hybrid form in the future, and research is currently being conducted in this field.

Finally, projection 21 focused on the scenario in which plant directors would be able to manage multiple factories centrally from their home office due to the complete and real-time transparency of all the operations in a digital system. In contrast to the shop floor level, management is more abstract. The processes that occur are not necessarily linked to and therefore dependent on physical objects. Management is therefore inherently more suited to teleworking, at least superficially. However, here too, the question must be asked as to what benefits and advantages the employee receives from teleworking. For example, a disadvantage is that personnel management, in particular, requires personal interaction. In this field, researchers agree that we will also end up with a hybrid solution. The experts interviewed did not consider the issue very relevant, as the group of people affected is small. They pointed out that some functions – especially in monitoring – are already possible. They also stated that personal contact and human-to-human interaction are important, especially as a manager.

It must be highlighted that several of the projections in this study have the potential to be used in the future: projections 18, 20, and (in parts) 21 show that it is possible to facilitate work and increase productivity at the same time. Furthermore, the attractiveness of workplaces can be increased with more flexible working hours and protection against infection or other dangers due to teleoperation.

Considering all the projections in this dimension, it becomes apparent that the HMI projections evoke various challenges at first glance. For a long time now, HMI has no longer consisted of buttons and switches only, but rather of incorporated digital displays, dashboards, and touch screens in modern control systems. It is possible to divide the challenges into general potential and barriers on a meta-level. As in all transformation processes, where interactions between humans and machines take place, it is essential to consider the humans' needs, at least initially, to increase acceptance (Hartson & Pyla, 2018). Acceptance is understood as the willingness to use or work with a specific type of interface. In essence, acceptance is largely determined by the ergonomics and usability of a technology. Thus, the mentioned aspects regarding the results of the Delphi study should be understood as indications for the successful implementation of a roadmap to the transformation into Next Generation Manufacturing.

Acknowledgment Funded by the Deutsche Forschungsgemeinschaft (DFG, German Research Foundation) under Germany's Excellence Strategy – EXC-2023 Internet of Production – 390621612.

References

Brauner, P., Dalibor, M., Jarke, M., Kunze, I., Koren, I., Lakemeyer, G., Liebenberg, M., Michael, J., Pennekamp, J., Quix, C., Rumpe, B., van der Aalst, W., Wehrle, K., Wortmann, A., & Ziefle, M. (2022). A computer science perspective on digital transformation in production. *ACM Transactions on Internet of Things, 3*(2), 1–32. https://doi.org/hg56

Brauner, P., Schaar, A. K., & Ziefle, M. (2022). Interfaces, interactions, and industry 4.0 — Why and how to design human-centered industrial user interfaces in the internet of production. In C. Röcker & S. Büttner (Eds.), *HTI – Shaping the future of industrial user interfaces*. Springer.

Brunner, O., Mertens, A., Nitsch, V., & Brandl, C. (2021). Accuracy of a markerless motion capture system for postural ergonomic risk assessment in occupational practice. *International Journal of Occupational Safety and Ergonomics*, 1–9. https://doi.org/hg5n

Dahlmanns, M., Lohmöller, J., Fink, I. B., Pennekamp, J., Wehrle, K., & Henze, M. (2020). Easing the conscience with OPC UA: An internet-wide study on insecure deployments. In *Proceedings of the ACM Internet Measurement Conference* (pp. 101–110) https://doi.org/hg5p

Deci, E. L., & Ryan, R. M. (2008). Self-determination theory: A macro theory of human motivation, development, and health. *Canadian Psychology, 49*(3), 182–185. https://doi.org/fcnwzz

Dix, A., Finlay, J., Abowd, G.D., & Beale, R. (Eds). (2003). Human computer interaction. Pearson.

Fayard, A. L., Weeks, J., & Khan, M. (2021). Designing the hybrid office. *Harvard Business Review, 99*(2), 114.

Garcia-Garcia, J. M., Penichet, V. M., & Lozano, M. D. (2017). Emotion detection: A technology review. In *Proceedings of the XVIII international conference on human computer interaction* (pp. 1–8) https://doi.org/gd6dwb

Gawer, A. (2014). Bridging differing perspectives on technological platforms: Toward an integrative framework. *Research Policy, 43*(7), 1239–1249. https://doi.org/gc8sc5

Gorecky, D., Schmitt, M., Loskyll, M., & Zühlke, D. (2014). Human-machine-interaction in the industry 4.0 era. In *2014 12th IEEE international conference on industrial informatics (INDIN)* (pp. 289–294) https://doi.org/ggz7vg

Hartson, R., & Pyla, P. S. (2018). *The UX book: Agile UX design for a quality user experience.* Morgan Kaufmann.

Hoff, K. A., & Bashir, M. (2015). Trust in automation: Integrating empirical evidence on factors that influence trust. *Human Factors, 57*(3), 407–434. https://doi.org/f68kpx

Kaber, D. B., & Endsley, M. R. (2004). The effects of level of automation and adaptive automation on human performance, situation awareness and workload in a dynamic control task. *Theoretical Issues in Ergonomics Science, 5*(2), 113–153. https://doi.org/10.1080/1463922021000054335

Lund, S., Madgavkar, A., Manyika, J., & Smit, S. (2020). *What's next for remote work: An analysis of 2,000 tasks, 800 jobs, and nine countries* (pp. 1–13). McKinsey Global Institute.

Papcun, P., Kajáti, E., & Koziorek, J. (2018). Human machine interface in concept of industry 4.0. In *2018 World Symposium on Digital Intelligence for Systems and Machines (DISA)* (pp. 289–296) https://doi.org/hg5s

Prewett, M. S., Johnson, R. C., Saboe, K. N., Elliott, L. R., & Coovert, M. D. (2010). Managing workload in human–robot interaction: A review of empirical studies. *Computers in Human Behavior, 26*(5), 840–856. https://doi.org/b2chxq

Schmidt, A. (2000). Implicit human computer interaction through context. *Personal Technologies, 4*, 191–199. https://doi.org/10.1007/BF01324126

Schumacher, S., Pokorni, B., Himmelstoß, H., & Bauernhansl, T. (2020). Conceptualization of a framework for the Design of Production Systems and Industrial Workplaces. *Procedia CIRP, 91*, 176–181. https://doi.org/hg5q

Schwab, K. (2017). The Fourth Industrial Revolution, The World Economic Forum, Foreign Affairs.

Siemens. (2020, July 16). Siemens to establish mobile working as core component of the "new normal" [Press Release]. https://assets.new.siemens.com/siemens/assets/api/uuid:1927cd36-a286-410a-a05f-4fa8acc35df9/HQCOPR202007155943EN.pdf

Valdeza, A. C., Braunera, P., Schaara, A. K., Holzingerb, A., & Zieflea, M. (2015). Reducing complexity with simplicity-usability methods for industry 4.0. In *Proceedings 19th triennial congress of the IEA* (Vol. 9, p. 14). https://doi.org/ggqvq6

Weck, M. (2006). *Werkzeugmaschinen 4: Automatisierung von Maschinen und Anlagen.* Springer-Verlag.

Wu, H., & Chen, Y. (2020). The impact of work from home (wfh) on workload and productivity in terms of different tasks and occupations. In *International Conference on Human-Computer Interaction* (pp. 693–706). Springer. https://doi.org/hg5t

Resilience Drivers in Next Generation Manufacturing

Alexander Schollemann, Marian Wiesch, Christian Brecher, and Günther Schuh

Abstract Given the high uncertainty in the likelihood of occurrence and the technical, economic, and societal impacts of digital transformations in the manufacturing industry, we conducted a technology foresight study, in the form of a real-time Delphi analysis, to derive reliable future scenarios featuring the next generation of manufacturing systems. This chapter presents the resilience dimension and describes each projection in detail, offering current case study examples and discussing related research, as well as implications for policy makers and firms. The current COVID-19 pandemic and its impact on human health, the biosphere on which we depend, and our need for certain commodities demonstrate the importance of developing global resilience. In 2030, supply chains are expected to be more decentralized, with production and sourcing moving closer to the end customer. Centralized production networks have been shown to be vulnerable to disruptions, and this trend is likely to continue. The majority of the experts do not expect production costs to rise substantially as a result of more regional production and higher inventory levels in order to cope with global crises. Some experts see reshoring, which is characterized by flexibility and resilience despite supposedly high costs in high-wage regions as a key long-term driver. In the future, production costs, while still important, will only be one factor taken into consideration by customers. The experts predict that AI-based decision-making systems will not be able to significantly increase production resilience by 2030. Factors such as lack of acceptance and the complexity of production networks are hindering the widespread implementation of such systems. However, companies that are already investing in AI see significant potential of this technology to help them overcome the challenges posed by global crises.

[Abstract generated by machine intelligence with GPT-3. No human intelligence applied.]

A. Schollemann (✉) · M. Wiesch · C. Brecher · G. Schuh
Laboratory for Machine Tools and Production Engineering (WZL), RWTH Aachen University, Aachen, Germany
e-mail: a.schollemann@wzl.rwth-aachen.de; m.wiesch@wzl.rwth-aachen.de; c.brecher@wzl.rwth-aachen.de; g.schuh@wzl.rwth-aachen.de

1 Introduction

The current COVID-19 pandemic and its impact on human health, the biosphere on which we depend, and our need for certain commodities demonstrate the importance of developing global resilience. As defined by sustainability scholars, resilience is the maintenance of development in the face of surprising and anticipated changes when there are thresholds between alternative paths, some of which are less desirable than others, and when it is difficult or even impossible to turn back once the threshold is crossed (Folke, 2006).

An analogy from biology with implications for human behavior can be used to illustrate the need to strengthen resilience, e.g., through novel methods, as a means of dealing with unexpected disturbances: "For example, a heatwave over a coral reef with low resilience, can result in algae taking over the corals. This often cause species loss, result in lower fish abundance, and large losses to the tourism sector. The algae are hard or impossible to get rid of, the corals might never come back. The people living off the reef can lose livelihoods dependent on tourism and fisheries. A resilient reef would be able to return back and the communities around it continue to develop. When a crisis hit, the resilience often comes from places less anticipated or even neglected. For example, a case study of a shift from algae to coral dominated communities on the Great Barrier Reef showed that recovery was primarily driven by one single species that is relatively rare on the reef, a species that easily can go extinct without notice under normal conditions but that is needed in crisis (Bellwood et al., 2006). Similarly, it is the less dominating species in a mature forest that often invades first and prepare the grounds for the return of the forest after a fire. Species that are less visible, and seem less valuable, can have a big role to play for the system to rebuild itself" (Gordon, 2020).

If we transfer this example to society during the COVID-19 pandemic, we can see how the importance of the work not only of doctors but also of nurses for the elderly has increased and how they have rebuilt the system or kept it alive. Transferring this example further, to the global industrial economy, the COVID-19 pandemic caused centralized supply chains, and thus entire production networks, to collapse in response to public health measures. The heavy reliance on Chinese suppliers, for example, meant that Apple had to curb production and at times allowed customers to buy only two iPhones per person.

To cope with this vulnerability to disruption, the importance of individual production systems within globally interconnected production networks must be assessed at each point in time so that production networks respond resiliently to disruptions. The increasing digitization of production, including the exchange of digital information, can be a significant technical and methodological enabler to increase resilience. In terms of the vulnerability of supply chains to disruptions, it will be important for companies to use operational data to make their supply chains more customized and independent.

Using a novel real-time Delphi approach (see chapter "Applying the Real-Time Delphi Method to Next Generation Manufacturing" for a presentation of the method and the sample, as well as chapter "Big Picture of Next Generation Manufacturing" for an overview of the results), we developed propositions for different scenarios within Next Generation Manufacturing in 2030. As suggested by Gawer (2014), we used an integrative framework for platforms, distinguishing four layers: governance (e.g., open forms of collaboration; see chapter "Governance Structures in Next Generation Manufacturing"), organization (e.g., boundaries and decision-making; see chapter "Organization Routines in Next Generation Manufacturing"), capabilities (e.g., hybrid intelligence; see chapter "Capability Configuration in Next Generation Manufacturing"), and interfaces (e.g., open APIs and human-machine interfaces; see chapter "Interface Design in Next Generation Manufacturing"). In addition, and influenced by our shared experiences during the COVID-19 pandemic, we added a fifth cluster of propositions addressing the need for resilience in future digital manufacturing systems (see this chapter). We provide a set of 24 validated projections based on 1930 quantitative estimations and 629 qualitative arguments from 35 industrial and academic experts from Europe, North America, and Asia. In so doing, we deliver a basis on which to substantiate academic discussions and which can support firm decision-making on future technological developments and economic implications that go beyond current speculations and siloed research.

Projections were developed to assess the ability of increasingly connected production networks to respond to unanticipated disruptions in a resilient manner. These projections were evaluated by the experts in terms of their probability of occurrence and their benefit in relation to the problem described (see Fig. 1).

Projection 22: *In 2030, supply chains will have become more decentralized with production and sourcing moving closer to the end customer to cope better with global crises* (e.g., pandemics).

Projection 23: *In 2030, production costs will have increased substantially due to more regional production and higher inventory levels to cope with global crises* (e.g., pandemics).

Projection 24: *In 2030, AI-based decision systems will enable greater resilience of production networks in the event of a global crisis* (e.g., a pandemic).

2   Projection 22: Decentralization

The first projection within the field of resilience drivers states that in 2030, supply chains will have become more decentralized, with production and sourcing moving closer to the end customer to cope better with global crises. The study estimates its probability of occurrence to be high, with a mean probability of 56.55%. Experts with an academic background and experts from outside of Germany consider the probability to be even higher (mean probability within academia, 63.00%; mean probability ROW, 63.75%). As the interquartile range (*IQR*) is low, with a value of 2.00, a consensus was found among the experts about this probability of occurrence. The expected influence on companies is considered relatively high, with a mean firm impact value of 3.17.

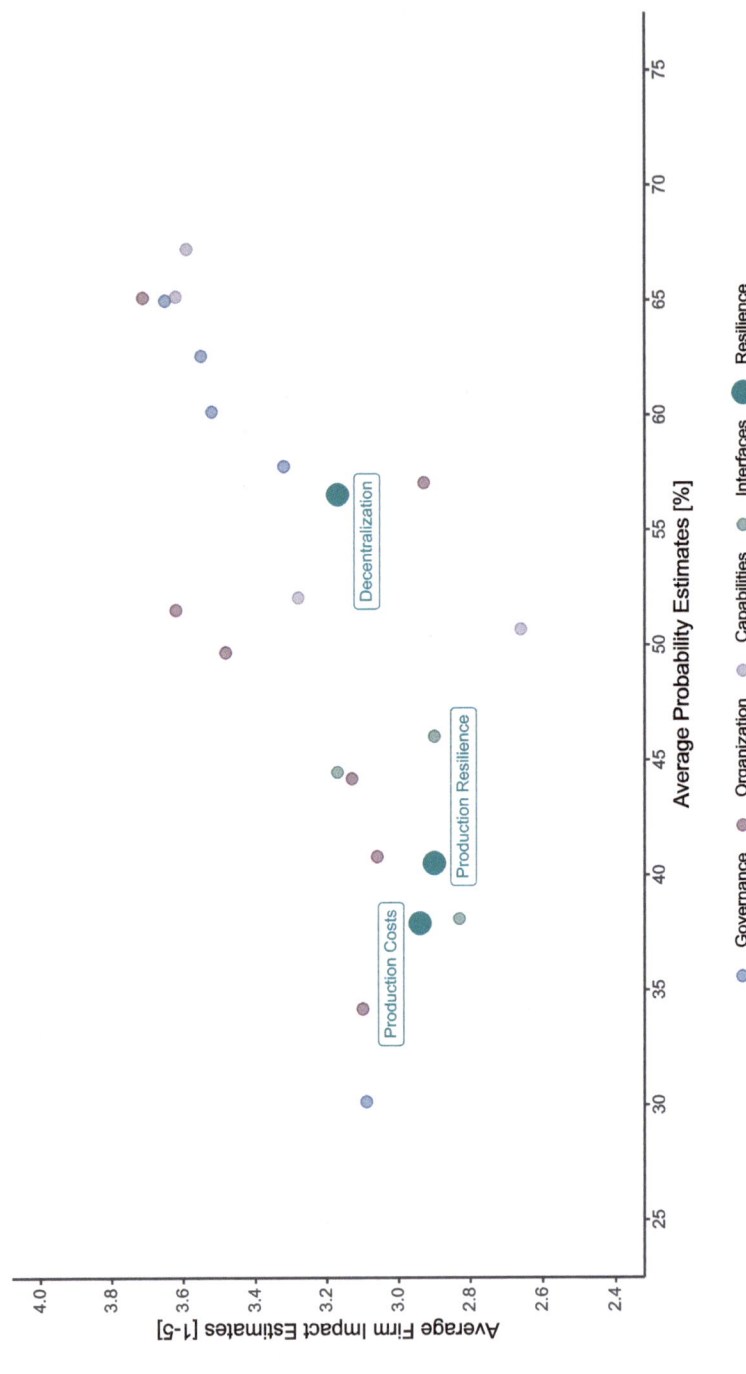

Fig. 1 Expert assessment on resilience drivers in Next Generation Manufacturing (see chapter "Big Picture of Next Generation Manufacturing" for the full results of the Delphi survey)

Recent years have clearly shown that global production networks are susceptible to disruptions, with the COVID-19 epidemic exposing the weaknesses of supply chains (Chowdhury et al., 2021). The high probability of occurrence and the high impact on companies of this projection can be explained by the impact on the economy seen in current events such as the blockade of the Suez Canal and the associated obstruction of global shipping traffic. Experts considering the probability to be lower argue that the current trend toward decentralization may abate once the COVID-19 crisis is over. Additionally, company size could be a decisive factor. However, decentralized production could offer a long-term solution approach, as current trends toward greater self-sufficiency and pressures to reduce carbon footprints, as well as rising geo-political trade conflicts, make its occurrence likely according to the experts. The experts indicate that the impact on the operational structure of companies will be medium if the projection occurs, mainly due to its impacts at the production network level.

During the COVID-19 pandemic, the vulnerability of centralized production networks became apparent. The heavy dependence on Chinese suppliers led to Apple having to cut back on production, and customers were only allowed to buy two iPhones per person at times. Honda also had no choice but to temporarily shut down its production site in Wuhan (Cai & Luo, 2020). Amazon closed a warehouse in Kentucky, and the food supplier Sysco had to reduce its workforce due to restrictions in the restaurant industry. As global crises, e.g., pandemics, may occur more frequently in the future, companies need to be prepared to withstand economic pressure. Decentralization can solve many of the associated problems, such as dependence on one country or its regulations. It will be important for companies to use operational data to make their supply chains more customized and independent. Society will benefit because it will be less vulnerable in times of crisis and purchasing power will generally increase. The challenge is to find a mix of risk avoidance and optimal use of data.

In this context, the data-based identification of adaptation needs in the design of global production networks represents a fundamental challenge. The adaptations required in response to the various changes mentioned are currently often made too late, however, because due to their complexity, immediate adaptation of network designs is not feasible (Moser et al., 2016). One of the main reasons is the delayed recognition of changes and thus of the need for adaptation. Accordingly, considering the hysteresis time in the design of global production networks, the latency period from the occurrence of the change to the perception of the change to the identification of the need for adaption must be shortened (Lanza et al., 2019). Recognition of changes in the design of production networks, analysis of possible consequences and correlations, and the derivation of network adaptation needs are requirements for the proactive identification of adaptation needs. Based on the systematization of the adaptation needs in the network, a data-based analysis of their causes and correlations can be performed by identifying those factors influencing KPIs used to monitor the network performance using a data-based approach. Finally, this is transferred into the derivation of adaptation needs in order to support companies in the future in network adaptation such as the decentralization of the global production network (Schuh et al., 2020).

3 Projection 23: Production Costs

The survey indicates a rather low mean probability of occurrence of 37.9% that production costs will substantially increase due to more regional production and higher inventory levels in order to cope with global crises such as pandemics. Experts from academia or from outside of Germany rate the probability of occurrence as higher (mean probability within academia, 48.00%; mean probability ROW, 50.00%). The same result emerges in the evaluation of the impact on companies if the projection occurs, which is assessed as medium overall, with a value of 2.94. The experts with academic backgrounds assess the impact on the company as high, with a firm impact factor of 3.50.

Most experts note that COVID-19 will indeed lead to a shift in supply chain strategies toward greater resilience, with regional production as one possible change. In this context, reshoring activities in global production networks are increasingly gaining significance as ways to accelerate the shift to regional production. However, these must be differentiated regarding their strategic orientation between short-term, reactive reshoring decisions and long-term strategic decisions of the overall supply chain (Barbieri et al., 2020). Some authors, therefore, see reshoring as a key long-term driver, characterized by flexibility and resilience despite supposedly high costs in high-wage regions. This is reinforced by new forms of automation, which have the potential to prevent significant cost increases. In addition, more resilient production could prove more cost-efficient in the long term, e.g., through fewer supply bottlenecks. Although reshoring is associated with higher production costs compared with the cost-oriented distribution of locations in global production networks, the majority of experts do not expect production costs to rise substantially. There is disagreement on the long-term trend, with some experts considering reshoring a trend that will not prevail in the face of the resumption of globalization.

In the future, production costs, while still important, will only be one factor taken into consideration by customers. Future generations will most likely not only evaluate products based on price but also consider social and environmental factors. The holistic consideration of sustainability dimensions is thus becoming increasingly important for companies. In this context, the sustainability dimensions are defined as economic, environmental, and social sustainability. For instance, the current gold standard for production efficiency, overall equipment effectiveness (OEE), is being adjusted to include sustainability factors (Boos, 2021). However, as well as addressing these aspects in their products, processes, and sites, manufacturing companies also have opportunities to exert an influence on production costs and thus on the associated dimensions of sustainability through the design of global production networks. Access to resources, e.g., green power or primary resources, and the regulations regarding environmental and social sustainability across globally distributed regions can have significant long-term impacts on production costs or on customers' purchase decisions.

The economic repercussions of the COVID-19 pandemic have been tremendous. Building an economy, and thus a production network, that is better able to cope with such shocks will be critical in the upcoming years. However, automation and intelligent use of data mean this will not necessarily lead to higher production costs. Even if production costs increase slightly, firms can still be competitive. Customers are increasingly evaluating products not solely based on cost measures but are also demanding environmental and social sustainability.

4   Projection 24: Production Resilience

The experts assume a low probability that in 2030, AI-based decision-making systems will enable greater resilience in production networks in the event of a global crisis (e.g., pandemic). The study estimates the probability of occurrence to be low, with a mean probability of 40.52%. As the interquartile range (IQR) is high, with a value of 2.90, disagreement was found among the experts about this probability of occurrence. This projection's influence on companies is considered to be medium, with a mean firm impact value of 2.90. The reason for this is that it will take longer than by 2030 to integrate AI systems so fundamentally into all areas that they can be used to cope with complex problems. Acceptance of such systems will also take time.

According to the experts, AI-based, all-encompassing decision-making systems for predicting the effects of and possible actions to be taken in response to unexpected events in order to increase resilience within production still require lengthy development due to their complexity, and their implementation would also be influenced by the need for acceptance of such systems. On the other hand, the experts concede that such systems, if acceptance can be demonstrated, have great leverage as a decision-making aid, since they are able to derive complex and unknown causalities implicitly contained in the totality of the data. These causalities can form the basis for important decisions, and thus the systems can be a major contribution to increasing resilience in production. Nevertheless, the acceptance of a holistic, AI-based decision-making system designed to respond to unexpected global events in production is highly dependent on how well it can predict new events, their effects, and possible responses using the knowledge it has learned so far.

Moreover, this acceptance would also depend on the comprehensibility of and the level of trust in the proposed decision, as well as in the handling of tangential issues like data protection, moral overlaps such as human surveillance, and legal responsibility in the event of wrong decisions. In addition, global crises always have new and unknown characteristics, which can mean that data patterns learned by AI from the past do not or only partially reappear, and therefore the trust level for and acceptance of AI-based decision systems will decrease. In particular, the interplay between possible effects of unknown events and the complexity of production as a holistic system consisting of production units, supply chains, organizational processes, different data interfaces, and different participants brings high complexity, which significantly increases the acceptance tolerance threshold.

For this reason, AI-based decision-making systems are not evaluated as a probable sole solution to dealing with global crises, but as an additional aid capable of transparently presenting previously unknown implicit relationships contained in historical data that may indicate causality, thereby assisting humans in dealing with a crisis.

AI tools already exist and are already used in production today. Capgemini (2019), Columbus (2020), McKinsey (2017), and McKinsey (2020) show concrete examples of how AI can be used in production and how it can be systematically scaled. The authors unanimously agree that AI is a game changer in manufacturing. It has the potential to change performance across the breadth and depth of manufacturing processes. Such tools produce quantitatively positive results, but often the user is not involved in finding the result. This black box leads to less acceptance. Coupled with the fact that production networks are very complex and significant projects, especially in crises, people do not want to place all responsibility into the hands of AI alone (Schuh et al., 2019). Despite the analyzed acceptance problem of AI, a global survey by McKinsey (2020) shows that following the economic challenges that pandemic response efforts brought to many companies, those who see the greatest benefit in AI are increasingly embracing the technology. These companies see a high potential in terms of increasing value creation through the use of AI and continue to invest in it, even during the pandemic. Most respondents from high-performing companies say their organizations have increased investment in AI across all key business areas in response to the pandemic. From the point of view of the leading companies in the study, AI as a future technology has the potential to deal with global crises. This could lead to a wider gap between the leading AI companies and the majority of companies that are still struggling to capitalize on the technology. It is necessary here to address and solve the problems mentioned in terms of acceptance of the use of AI applications in production. In contrast to the companies in McKinsey's (2020) study, the experts surveyed do not currently believe that AI in production is a comprehensive technology for dealing with crises. The leading companies in terms of investment levels in AI listed in McKinsey's survey demonstrate several practices that could provide helpful hints for success and thus close the gap in AI acceptance.

5   Summary

In this chapter, macroscopic effects were highlighted in the context of global crisis situations. In this respect, the COVID-19 pandemic in particular has highlighted the fragility of global production networks and their associated supply chains, which have been significantly shaped by the ongoing globalization of recent decades. Due to this thematic topicality, the present Delphi study investigated whether supply chains will become more decentralized in the next decade (P22) and whether production costs will increase significantly due to more regional production and higher inventories (P23). In addition, the study examined the potential of AI-based decision systems to enable greater resilience in production networks (P24).

Projection 22 (decentralization): the COVID-19 epidemic has exposed weaknesses in supply chains, which will lead to more decentralized production networks in the coming years, with production and sourcing moving closer to the end customer to cope better with global crises. In this context, data-based identification of required adaptions to the global production network design plays an important role in transparently detecting various risk factors and initiating appropriate measures along the production network.

Projection 23 (production costs): we found that despite the shift in supply chain strategies toward regional production triggered by the COVID-19 pandemic, this more resilient and flexible design of the production network is not necessarily accompanied by higher production costs. Through automation, the intelligent use of data, and, especially, a new awareness of environmental and social sustainability among customers, production costs will not play a substantial role.

Projection 24 (production resilience): we found that AI-based decision-making systems are not evaluated as a probable sole solution to dealing with global crises but could be an additional aid capable of transparently presenting previously unknown implicit relationships contained in historical data that may indicate causality, thereby assisting humans in dealing with the crisis. For many experts, the traceability of AI-based decisions is currently a black box, which is affecting acceptance. Nevertheless, other studies show that large companies, especially those that have experience with AI, are increasingly relying on this technology, as it is said to have great potential for increasing value creation, which increases their belief in their ability to better handle global crises.

This chapter investigated from the experts' perspectives how the vulnerabilities in global supply chains exposed by the COVID-19 pandemic will evolve within the long-term design of production networks. In this context, short-term emergency reactions such as the drastic ramping up of inventories are contrary to the long-term development of resilient network structures. Whether the seemingly unstoppable globalization drive will continue despite pandemics, rising protectionism, and global crises is yet to be critically questioned. Consistent reshoring coupled with the use of automation and digitization can enable high-wage locations to continue to operate economically while simultaneously increasing network resilience. In this context, however, it is unlikely that there will be a wave of reshoring, but rather a targeted design of the production network in which data-based, early detection of adaptation needs can provide transparency in the decision-making process.

Acknowledgment Funded by the Deutsche Forschungsgemeinschaft (DFG, German Research Foundation) under Germany's Excellence Strategy – EXC-2023 Internet of Production – 390621612.

References

Barbieri, P., Boffelli, A., Elia, S., Fratocchi, L., Kalchschmidt, M., & Samson, D. (2020). What can we learn about reshoring after COVID-19? *Operations Management Research, 13*(3), 131–136. https://doi.org/10.1007/s12063-020-00160-1.

Bellwood, D. R., Hoey, A. S., Ackerman, J. L., & Depczynski, M. (2006). Coral bleaching, reef fish community phase shifts and the resilience of coral reefs. *Global Change Biology, 12*(9), 1587–1594. https://doi.org/10.1111/j.1365-2486.2006.01204.x.

Boos, W. (2021). Production turnaround—Turning data into sustainability. Through the Internet of Production towards sustainable production and operation. Laboratory for Machine Tools and Production Engineering (WZL) of RWTH Aachen University, Aachen.

Cai, M., & Luo, J. (2020). Influence of COVID-19 on manufacturing industry and corresponding countermeasures from supply chain perspective. *Journal of Shanghai Jiaotong University (Science), 25*(4), 409–416. https://doi.org/10.1007/s12204-020-2206-z.

Capgemini. (2019). *Scaling AI in manufacturing operations: A practitioners' perspective.* Capgemini Research Institute.

Chowdhury, P., Paul, S. K., Kaisar, S., & Moktadir, M. A. (2021). COVID-19 pandemic related supply chain studies: A systematic review. *Transportation Research Part E: Logistics and Transportation Review, 148*, 102271. https://doi.org/10.1016/j.tre.2021.102271.

Columbus, L. (2020). 10 Ways AI is improving manufacturing in 2020. In Forbes Magazine.

Folke, C. (2006). Resilience: The emergence of a perspective for social–ecological systems analyses. *Global Environmental Change, 16*(3), 253–267. https://doi.org/10.1016/j.gloenvcha.2006.04.002.

Gawer, A. (2014). Bridging differing perspectives on technological platforms: Toward an integrative framework. *Research Policy, 43*(7), 1239–1249. https://doi.org/10.1016/j.respol.2014.03.006.

Gordon, L. J. (2020). The Covid-19 pandemic stress the need to build resilient production ecosystems. In Agriculture and Human Values (37, 3, 645–646). Springer Science and Business Media LLC. https://doi.org/10.1007/s10460-020-10105-w.

Lanza, G., Ferdows, K., Kara, S., Mourtzis, D., Schuh, G., Váncza, J., Wang, L., & Wiendahl, H.-P. (2019). Global production networks: Design and operation. *CIRP Annals, 68*(2), 823–841. https://doi.org/10.1016/j.cirp.2019.05.008.

McKinsey. (2017). *Smartening up with artificial intelligence (AI)—What's in it for Germany and its industrial sector?* McKinsey & Company.

McKinsey. (2020). *Global survey: The state of AI in 2020.* McKinsey & Company.

Moser, E., Stricker, N., & Lanza, G. (2016). Risk efficient migration strategies for global production networks. *Procedia CIRP, 57*, 104–109. https://doi.org/10.1016/j.procir.2016.11.019.

Schuh, G., Gützlaff, A., Thomas, K., & Rodemann, N. (2020). Framework for the proactive identification of adaptation needs in the configuration of global production networks. In 2020 IEEE International Conference on Industrial Engineering and Engineering Management (IEEM) (pp. 69–73). https://doi.org/10.1109/IEEM45057.2020.9309963.

Schuh, G., Prote, J. P., Sauermann, F., & Schmitz, S. (2019). Production analytics. *Zeitschrift für wirtschaftlichen Fabrikbetrieb, 114*(9), 588–591. https://doi.org/10.3139/104.112153.

Future Scenarios and the Most Probable Future for Next Generation Manufacturing

Marc Van Dyck, Sebastian Pütz, Alexander Mertens, Dirk Lüttgens, Verena Nitsch, and Frank T. Piller

Abstract Based on the results of a rigorous Delphi study, we present scenarios that portray a most probable future of Next Generation Manufacturing in 2030, enabled by connected data (digital shadows) shared in cross-organizational data spaces. We provide individual scenarios for the dimensions *governance*, *organization*, *capabilities*, *interfaces*, and *resilience*, as well as one aggregated scenario for the future development of the manufacturing ecosystem. Our analysis identifies two fundamental changes: a shift from the current focus in many Industry 4.0 use cases on operational efficiency toward more ecologically and socially sustainable production and an anthropocentric perspective complementing techno-centric production. We discuss emerging tensions resulting from these changes.

[Abstract generated by machine intelligence with GPT-3. No human intelligence applied.]

M. Van Dyck (✉) · D. Lüttgens
Institute for Technology and Innovation Management, RWTH Aachen University, Aachen, Germany
e-mail: vandyck@time.rwth-aachen.de; luettgens@time.rwth-aachen.de

S. Pütz · A. Mertens
Institute of Industrial Engineering and Ergonomics, RWTH Aachen University, Aachen, Germany
e-mail: s.puetz@iaw.rwth-aachen.de; a.mertens@iaw.rwth-aachen.de

V. Nitsch
Institute of Industrial Engineering and Ergonomics, RWTH Aachen University, Aachen, Germany

Fraunhofer Institute for Communication, Information Processing and Ergonomics FKIE, Aachen, Germany
e-mail: v.nitsch@iaw.rwth-aachen.de

F. T. Piller
Institute for Technology and Innovation Management, RWTH Aachen University, Aachen, Germany

Institute for Business Cybernetics (IfU) e.V., RWTH Aachen University, Aachen, Germany
e-mail: piller@time.rwth-aachen.de

F. T. Piller et al. (eds.), *Forecasting Next Generation Manufacturing*, Contributions to Management Science, https://doi.org/10.1007/978-3-031-07734-0_9

1   Overview

The way products are developed, produced, and distributed will fundamentally change in the next decade. To portray a most probable future of Next Generation Manufacturing in 2030, this chapter presents multiple scenarios which focus on different dimensions in the framework suggested by Gawer (2014) and introduced in more detail in chapter "How Digital Shadows, New Forms of Human-Machine Collaboration, and Data-Driven Business Models Are Driving the Future of Industry 4.0": *Governance* (e.g., open forms of collaboration), *organization* (e.g., boundaries and decision-making), *capabilities* (e.g., hybrid intelligence), *interfaces* (e.g., open APIs and human-machine interfaces), and *resilience* (e.g., decentralization). The scenarios are substantiated by the collected quantitative (see chapter "Big Picture of Next Generation Manufacturing") and qualitative findings of the conducted Delphi study and highlight those projections that were estimated as most likely and most impactful. Direct quotes from the experts' responses are used for illustrative purposes and to emphasize the role of these scenarios as syntheses of the collected expert assessments. Figure 1 summarizes a most probable scenario for Next Generation Manufacturing organized along the five dimensions and from internal and external perspectives.

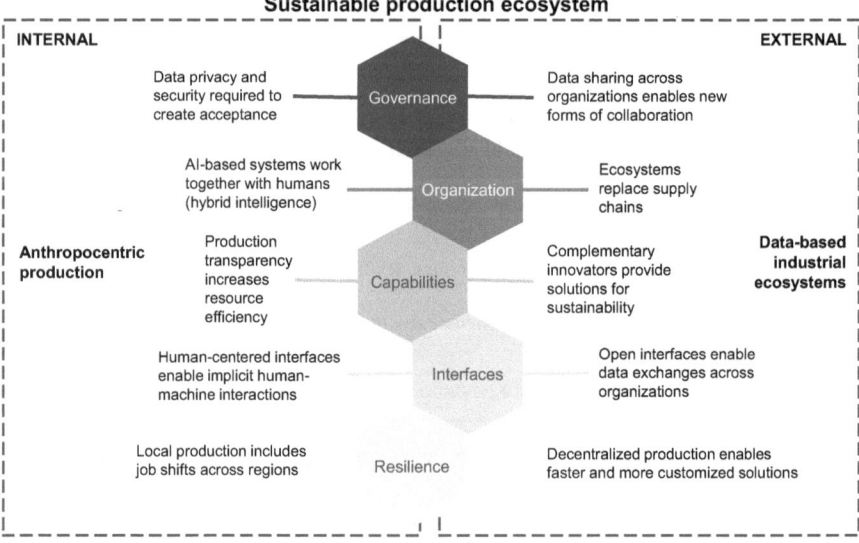

Fig. 1 Most probable scenario for Next Generation Manufacturing

2 Governance

The projections that form the most relevant scenario for the *governance* dimension for the year 2030 address *subscription models* (P1), *digital services* (P2), *data sharing* (P3), and *industrial GDPR* (P6):

Scenario The introduction of digital shadows creates opportunities for business models with new forms of collaboration between suppliers, manufacturers, and customers. Increasingly, customers want to "buy guaranteed production capabilities and not just machines." Subscription models where customers only pay for the use or outcome of a machine are also a potential catalyst for new technologies. In particular, highly specialized equipment could benefit from more flexible models which change investments from capital expenditures (CAPEX) to operating expenses (OPEX). Meanwhile, existing legacy machines that are focused on producing at high volume will likely remain in traditional ownership models with a clear depreciation calculation. Here, competitive advantage and differentiation will be achieved "only via software," as physical efficiencies are exhausted. Since "many firms lack the data capabilities" required to create software-based efficiencies by leveraging data from digital shadows, machinery suppliers will offer them as digital services and benefit from data learning effects across their customers. In contrast, for specialized equipment, hardware remains crucial, and a "return to much more competition based on hardware capabilities due to innovation in materials, design, functionality, and shrinking global supply chains" is expected.

A fundamental requirement for this scenario is data sharing across organizations, i.e., digital shadows that incorporate usage data from various customers and make supply chains transparent. While the traceability of components offers clear benefits in terms of compliancy with regulations such as the Act on Corporate Due Diligence Obligations in Supply Chains (BMAS, 2021) and opens up opportunities for efficiency gains, many firms are reluctant to share data. Internally, employees and managers fear security and privacy issues given the extensive data collection required, which may violate ethical requirements. Externally, firms fear exposing firm secrets when sharing data with other parties in the supply chain: "manufacturers want to keep control." At the same time, small- and medium-sized firms feel forced by large players to share data despite their doubts.

Creating acceptance by all stakeholders and ensuring security and privacy will be key for these new collaboration modes that require data sharing and involve joint risk-taking. Otherwise, potential negative effects, such as monopoly building due to data network effects or failure of new business models due to reluctance to share data, will prevail (Piller et al., 2021).

(continued)

However, it is still unclear how this acceptance can be achieved. One avenue is new anonymization methods that allow secure data sharing without compromising firm secrets and which ensure privacy protection. Another avenue is new data protection regulations for industrial sectors, equivalent to the European General Data Protection Regulation (European Parliament and Council, 2016) for consumers. While this provides an opportunity for the EU to set guidelines with potentially global impacts, it is unlikely to be realized before 2030. Until then, data sharing agreements will be managed in bilateral contracts. However, this involves a risk that proprietary standards managing data sharing could be established by powerful players to the detriment of smaller firms. Regulators should watch this closely.

3   Organization

The projections that form the most relevant scenario for the *organization* dimension for the year 2030 address *hybrid intelligence* (P8), *AI assistants* (P9), *employees' rights* (P12), and *workforce reduction* (P13):

Scenario The development of the industrial internet along with the introduction of digital shadows will shape decision-making processes in manufacturing companies across all organizational levels. As time progresses, the role of AI-based decision support systems will continue to grow and will have reached substantial level by 2030. One important use case for decision support systems will be the support of managers, as "AI and statistics can help to prepare data for strategic decision-making." Thereby, the application of AI-based algorithms will facilitate decisions "based on facts and data," reducing the influence of managers' gut feelings and intuitive judgments. Moreover, decision support systems will make it much easier for managers "to consider a multitude of influencing factors simultaneously and perform multi-criteria optimization." However, by 2030, the support of AI-based applications may still be limited to "short-term decision-making," and their use for long-term decision-making on a strategic level may only be achieved at a later stage. The main obstacle in this regard is the scarcity of adequate data, resulting in a high level of uncertainty even for well-calibrated algorithms.

In line with their use at the managerial level, AI-based decision-making agents will play a major role in operative production decisions. Here, automated decision-making will be particularly useful because operative decisions are mostly "repetitive and a large data amount is available," so they can be used "for decision templates or scenario development." Nonetheless, AI-based decisions will still need to be supervised by human actors who will retain

(continued)

"final decision-making power," upholding the relevance of human expertise. Consequently, the introduction of AI-based decision systems will not lead to a drastic reduction of the total workforce. Whereas the number of shop floor workers might be reduced by about 20% due to industrial robots taking over "repetitive tasks," new "supportive jobs" for managing the AI-based systems will emerge. For companies, it will be crucial to hire or train workers with the appropriate digital skills to capitalize on the potential competitive advantages of using AI technologies. Regarding the data sources of the digital shadows that form the basis for the AI-based decision support systems, most implementations in 2030 will access data on production machinery and process parameters. In contrast, digital shadows of production workers will only achieve limited application due to the associated risks of invading employees' privacy. For those applications implemented in practice, unions will enforce "advanced anonymization techniques [as] a prerequisite" to prevent individual workers from being identifiable, reducing the potential for misuse of the data.

4   Capabilities

The projections that form the most relevant scenario for the *capabilities* dimension for the year 2030 address *environmental sustainability* (P15) and *production transparency* (P16):

Scenario Digital shadows serve as a key enabler of both producing and developing more environmentally sustainable products. Environmental sustainability will become a decisive competitive factor demanded by customers and employees alike and "enforced by financial markets." It may become central to a new corporate culture and boost employee morale, thanks to a "higher purpose." While all interests seem to be aligned, the extent of this change will vary depending on the industry and the country. From a global perspective, it "largely depends on what happens in emerging countries," which are more financially constrained and already suffering from the effects of a changing climate. From an industry perspective, operating in an environmentally sustainable manner is more difficult for asset-heavy industries and certain technologies.

Even though there is "no other choice" but to comply with the need for environmental sustainability, it is unclear which technologies will help achieve this. In principle, transparency and data availability support sustainability. They enable better forecasting, scenario analyses, and a reduction of waste, potentially leading to an efficiency increase of 10–30%. However, production transparency will not be fully realized by 2030 due to a lack of data readiness

(continued)

and a lack of digital competencies among small- and medium-sized firms. Furthermore, other approaches are competing with digital shadows, such as transparency through blockchain or sustainability certificates issued by independent authorities.

5 Interfaces

The projections that form the most relevant scenario for the *interfaces* dimension for the year 2030 address *implicit interfaces* (P18) and *open data interfaces* (P19):

Scenario While the implementation of digital shadows will undoubtedly require the introduction of new interfaces both within production systems and between companies, the potential advances will not be fully realized by 2030. Starting with human-machine interfaces on the shop floor, increasing levels of automation and the application of AI-based support systems will further reduce the amount of manual work performed by human actors and shift their tasks toward collaboration with robots, automation control, automation supervision, and decision selection. Consequently, production workers will increasingly be confronted with digital interfaces that provide insights into or from AI-based systems. In terms of optimally designing these interfaces between human actors and AI-powered automation, a promising approach is to move from explicit interactions, with the human operator exercising direct control, to implicit interactions in which the machine automatically adapts to human behavior by recognizing and predicting human activities. The deployment of AI in this context offers a multitude of opportunities because it "will enable systems to prepare and make better decisions." Furthermore, the required "sensor technologies are developing fast and [are] getting cheap," as is the hardware required for the necessary "processing power." However, "production systems [are] slow to adapt," so the timeframe under consideration, up to 2030, "is too early." In addition, the adoption of implicit interfaces will require high standards for robustness and reliability. Beyond 2030, implicit interactions between production workers and AI-based systems will be pursued, although a status of "100% implicit interaction" will not be targeted.

Regarding the new demands for external interfaces associated with Next Generation Manufacturing, the exchange of data between companies will require standardized interfaces to enable efficient exchange processes and new business models. Here, the introduction of regulatory requirements demanding open and standardized interfaces "would be a major game changer," as they would "boost the [development of] standards." However,

(continued)

such standards are rather "unlikely in [the] next 10 years." Standardization efforts will be hampered by companies relying on their "installed base and legacy" production machines. In addition, "different manufacturers [will] still protect their competitive knowledge and data and [will] resist full integration/ connectivity." Therefore, regulations for standardized data interfaces are likely to be implemented only after 2030, with the first regulations likely to be enacted for an "economic bloc like the EU" rather than globally.

6 Resilience

The projection that forms the most relevant scenario for the *resilience* dimension for the year 2030 addresses *decentralization* (P22):

Scenario The COVID-19 pandemic caused significant supply chain disruptions and created supply and labor shortages. Better AI-based decision support systems and forecasting mechanisms will only have a minor impact in dealing with future crises, as such events are by definition outliers which cannot be predicted easily by learning from the past. Instead, more flexible production setups are required. Decentralization of production is a viable option that allows production to move closer to the customer, thereby making delivery faster and solutions more customized. In this scenario, production plants will not necessarily produce exclusively for one manufacturer. "Manufacturing-as-a-service," where plant operators provide production capacity to multiple firms, thanks to a more flexible and interconnected production setup based on digital shadows, offers new opportunities. "Trends toward de-globalization, greater self-sufficiency, and pressures to reduce carbon footprint, together with rising geo-political trade conflicts," will exacerbate regionalization trends. This also includes a job shift across regions. However, while the current focus is on creating robust and resilient supply chains for future pandemics, experts are wary as to whether this trend will reverse to globally centralized production after COVID-19 due to cost efficiency.

7 A Most Probable Scenario for Next Generation Manufacturing

A scaled deployment of digital shadows that connect data, products, and equipment across organizational boundaries would significantly change the way products are developed, produced, and distributed. Our Delphi study showcases how these changes will materialize in a vision of Next Generation Manufacturing by 2030.

Two fundamental changes lie at the center of Next Generation Manufacturing: first, a shift from operational efficiency to a broader set of economic, ecological, and social sustainability principles in production, and second, an anthropocentric perspective on production, where machines learn from humans and vice versa.

Scenario Digital shadows enable cross-company data spaces which create transparency along the value chain and a product's life cycle (Otto & Jarke, 2019; Cappiello et al., 2020). In combination with machine learning and simulation models, new insights and collaboration beyond organizational boundaries open up new opportunities for sustainable production (Piller et al., 2022). For instance, production equipment can be monitored throughout its lifetime to improve its efficiency and longevity through predictive maintenance and to enable second life use cases or recycling through exact health measures.

Internally, digital shadows offer the opportunity to complement a techno-centric understanding of production with an anthropocentric perspective (Mertens et al., 2021). Human shop floor workers and production management will continue to play key roles in manufacturing, but the type of work and the required skills will change. In Next Generation Manufacturing, machines and humans will be much more interconnected. Artificial intelligence-based decision support systems will complement human decision-making for tactical and operational tasks. In addition, new forms of human-machine collaboration will require more emphasis on user-centered machine interfaces.

Externally, cross-company data spaces enabled by digital shadows will create the foundation for data-based industrial ecosystems. Decentralized manufacturing systems will allow flexible and local production to be more resilient and closer to the customer. At the same time, these decentralized systems will be connected through open interfaces that enable data exchanges between digital shadows in (nearly) real time. The data generated in these ecosystems will provide new opportunities for differentiation while changing the forms of collaboration. Third parties will be able to tap into this data to develop service applications that help achieve greater efficiency and sustainability. This resembles platform-based business models known from consumer settings (Kopalle et al., 2020). Consequently, industry structures will change where manufacturing firms can take on new roles, such as app complementors or platform orchestrators (Piller et al., 2021). However, these forms of value creation and capture require data sharing beyond organizational boundaries. To ensure privacy and data security, new regulations for data protection will need to be in place.

8 Summary: Tensions Arising at the Interplay Between Internal and External Perspectives

The above scenarios illustrate the most probable scenarios with a high impact on employees, managers, firms, and society. Nevertheless, the impact of digital shadows is controversial and uncertain, which is common for technology-related Delphi studies (e.g., Jiang et al., 2017). One reason for uncertainty is the latent tensions between the company-internal and the external (network) perspectives. In particular, the governance and organization dimensions require a delicate balance between the two perspectives. For instance, firms need to manage the need for internal data privacy and protection and the need for external data sharing. Furthermore, firms are confronted with a lack of data science skills among their employees and a lack of acceptance of AI-based decision support systems despite the need for faster, automated, and data-based decision-making across organizations. These tensions lie at the fundament of how digital shadows will be implemented and how they will affect manufacturing firms (Bauernhansl et al., 2018). Depending on how managers, customers, and regulators manage these tensions, the scenarios will materialize differently.

Acknowledgment Funded by the Deutsche Forschungsgemeinschaft (DFG, German Research Foundation) under Germany's Excellence Strategy—EXC-2023 Internet of Production—390621612.

References

Bauernhansl, T., Hartleif, S., & Felix, T. (2018). The Digital Shadow of production: A concept for the effective and efficient information supply in dynamic industrial environments. *Procedia CIRP, 72*, 69–74. https://doi.org/gfgvgs

BMAS. (2021). *Act on corporate due diligence obligations in supply chains.* https://www.bmas.de/SharedDocs/Downloads/DE/Internationales/act-corporate-due-diligence-obligations-supply-chains.pdf;jsessionid=FEAA2DC9AA9ED79C9C23624381AAE27B.delivery1-master?__blob=publicationFile&v=3

Cappiello, C., Gal, A., Jarke, M., & Rehof, J. (2020). Data ecosystems: Sovereign data exchange among organizations (Dagstuhl Seminar 19391). *Dagstuhl Reports, 9*(9), 66–134. https://doi.org/hhk5

European Parliament and Council. (2016). *Regulation (EU) 2016/679 of the European Parliament and of the Council of 27 April 2016 on the protection of natural persons with regard to the processing of personal data and on the free movement of such data, and repealing Directive 95/46/EC (General Data Protection Regulation).* https://eur-lex.europa.eu/legal-content/EN/TXT/PDF/?uri=CELEX:32016R0679&from=EN

Gawer, A. (2014). Bridging differing perspectives on technological platforms: Toward an integrative framework. *Research Policy, 43*(7), 1239–1249. https://doi.org/gc8sc5

Jiang, R., Kleer, R., & Piller, F. T. (2017). Predicting the future of additive manufacturing: A Delphi study on economic and societal implications of 3D printing for 2030. *Technological Forecasting and Social Change, 117*, 84–97. https://doi.org/ghzgpp

Kopalle, P. K., Kumar, V., & Subramaniam, M. (2020). How legacy firms can embrace the digital ecosystem via digital customer orientation. *Journal of the Academy of Marketing Science, 48*(1), 114–131. https://doi.org/gj27r5

Mertens, A., Pütz, S., Brauner, P., Brillowski, F., Buczak, N., Dammers, H., et al. (2021, July). Human digital shadow: Data-based modeling of users and usage in the internet of production. In *2021 14th International Conference on Human System Interaction (HSI)* (pp. 1–8). https://doi.org/hg6g

Otto, B., & Jarke, M. (2019). Designing a multi-sided data platform: Findings from the International Data Spaces case. *Electronic Markets, 29*(4), 561–580. https://doi.org/ggqvq9

Piller, F. T., Van Dyck, M., Lüttgens, D., & Diener, K. (2021). Positioning strategies in emerging industrial ecosystems for industry 4.0. In *Proceedings of the 54th Hawaii International Conference on System Sciences* (pp. 6153–6163). https://doi.org/hgz3

Piller, F. T., et al. (2022). *Industry 4.0 and sustainability: How digital business models foster sustainability in industry*. In Position paper of the German Stakeholder Platform Industrie 4.0. Berlin: 2020. https://doi.org/hhk6

Hybrid Intelligence in Next Generation Manufacturing: An Outlook on New Forms of Collaboration Between Human and Algorithmic Decision-Makers in the Factory of the Future

Frank T. Piller, Verena Nitsch, and Wil van der Aalst

Abstract The text discusses the concept of hybrid intelligence, which is a form of collaboration between machines and humans. It describes how this concept can be used in manufacturing to help improve productivity. The text also discusses how this concept can be used to help humans learn from machines. There is a debate in the intelligence community about the role of humans vs. machines. Machine intelligence can do some things better than humans, such as processing large amounts of data, but is not good at tasks that require common sense or empathy. Augmented intelligence emphasizes the assistive role of machine intelligence, while hybrid intelligence posits that humans and machines are part of a common loop, where they adapt to and collaborate with each other. The text discusses the implications of increasing machine involvement in organizational decision-making, specifically mentioning two challenges: negative effects on human behavior and flaws in machine decision-making. It argues that, in order for machine intelligence to improve decision-making processes, humans and machines must collaborate. The chapter argues that hybrid intelligence is the most likely scenario for decision-making in the future factory. The chapter discusses the advantages of this approach and how it can be used to improve quality control in a production system. The transformer-based

F. T. Piller (✉)
Institute for Technology and Innovation Management, RWTH Aachen University, Aachen, Germany
e-mail: piller@time.rwth-aachen.de

V. Nitsch
Institute of Industrial Engineering and Ergonomics, RWTH Aachen University, Aachen, Germany

Fraunhofer Institute for Communication, Information Processing and Ergonomics FKIE, Wachtberg, Germany
e-mail: v.nitsch@iaw.rwth-aachen.de

W. van der Aalst
Chair of Process and Data Science, RWTH Aachen University, Aachen, Germany
e-mail: wvdaalst@pads.rwth-aachen.de

F. T. Piller et al. (eds.), *Forecasting Next Generation Manufacturing*, Contributions to Management Science, https://doi.org/10.1007/978-3-031-07734-0_10

language model called GPT-3 can be used to generate summaries of text. This task is difficult for machines because they have to understand sentiment and meaning in textual data. The model is also a "few-shot learner," which means that it is able to generate a text based on a limited amount of examples. Transformer-based language models are beneficial because they are able to take the context of the processed words into consideration. This allows for a more nuanced understanding of related words and concepts within a given text.

[Abstract generated by machine intelligence with GPT-3. No human intelligence applied.]

1 From Human-Computer Interaction to Human-Machine Collaboration

As summarized in the previous chapter "Future Scenarios and the Most Probable Future for Next Generation Manufacturing", the most probable scenario resulting from our Delphi study on the future of digitalization in manufacturing predicts two fundamental changes until 2030 that will be enabled by the scaled deployment of digital shadows connecting data, products, and equipment across organizational boundaries: first, a shift from the current focus on operational efficiency to a broader set of economic, ecological, and social sustainability objectives driving future manufacturing strategies and second, an anthropocentric perspective on production where machines learn from humans and humans from machines in a much more collaborative form as compared to the status quo today.

In this final chapter of our book, we build on the second development. It corresponds to the paradigm shift from a technology-centered toward a human-centered digitalization and work design, consistently reconsidering the role of humans in the factory of the future (Mütze-Niewöhner et al., 2022; Hirsch-Kreinsen & Ittermann, 2021). Chapters "Organization Routines in Next Generation Manufacturing" and "Capability Configuration in Next Generation Manufacturing" already discussed these developments in larger detail. Human-centered digitalization and work design are also a central element of our understanding of an "Industry 4. U," as introduced in the first chapter of this book, describing the next evolution of Industry 4.0—centered on people *and* planet.

Human-centered digitalization starts with using technology to support humans at work in an individually customized manner by taking individual capabilities, habits, and preferences into account. Nevertheless, it also has a profound impact on how decisions are made in an organizational context, enabled by new forms of collaboration between humans and machines (machine intelligence). Delphi Projection P8 proposed the rise of a "hybrid intelligence," suggesting that in 2030, "strategic production decisions will be executed in close interaction between humans and AI-based algorithms." Our expert panel demonstrated consensus and a high probability that this projection will be realized within the next decade.

In this chapter, we explore the concept of hybrid intelligence in larger detail. While there are more questions than answers and we are just at the beginning to investigate this concept, early examples are already here. We used a specific use case of (a weak) hybrid intelligence to write this book: a (transformer) language model helped us to compose the abstracts and summaries of this book. While probably just a simple form of hybrid intelligence, it still provides a good illustration of a new form of collaboration between machines and us. We will discuss this specific application and its technical background in larger detail toward the end of this chapter. Before, we outline our understanding and definition of hybrid intelligence and the open research questions it poses with regard to the future organization of work. In this context, we present a specific scenario of using hybrid intelligence for learning and continuous improvement for Next Generation Manufacturing.

2 Hybrid Intelligence: Concept and Definition

There used to be a clear separation between tasks done by machines and tasks done by people (van der Aalst, 2021). Machine intelligence, i.e., mixtures of artificial intelligence (AI) and machine learning (ML), can deal amazingly well with unstructured data (text, images, and video) as long as there are enough training data. In the corporate context, the use of machine intelligence attempts to make structures and processes more efficient. Applications in speech recognition (e.g., Alexa and Siri), image recognition, automated translation, autonomous driving, and medical diagnosis have blurred the classical divide between human tasks and machine tasks. However, while machine intelligence works well for such clearly defined tasks, it is not foreseeable that it will become capable of fully mapping complex business problems in organizational contexts (Dellermann et al., 2019) or solving multiple tasks simultaneously (Raj & Seamans, 2019). Although current AI and ML technologies outperform humans in many areas, tasks requiring common sense, contextual knowledge, creativity, adaptivity, or empathy are still best performed by human intelligence. Machine intelligence, on the contrary, is about data and algorithms and can be characterized by terms such as fast, efficient, cheap, scalable, and consistent.
    Taken together, Dellermann et al. (2019) define *hybrid intelligence* as:

> the ability to achieve complex goals by combining human and artificial intelligence, thereby reaching superior results to those each of them could have accomplished separately, and continuously improve by learning from each other.

Following this definition, hybrid intelligence hence blends human intelligence and machine intelligence to combine the best of both worlds. As things stand today, it is the most likely deployment scenario of machine intelligence in the corporate context over the next few decades. Hybrid intelligence aims to leverage the complementary strengths of human and machine intelligence in such a way that better overall performance can be achieved than when machines or humans are used alone (Dellermann et al., 2019; Kamar, 2016). Even in often-cited application scenarios that use AI-based algorithms for decision preparation or outsource decision-making to AI (e.g., laboratory data interpretation, human resources, claims processing), human actors invariably play a central role (Shrestha et al., 2019).

A closely related term, *augmented intelligence*, emphasizes the assistive role of machine intelligence (especially ML), when deep neural nets and other data-driven techniques enhance human intelligence rather than replace it. In this understanding, AI and ML are shifting human intelligence on a higher level, just like telescopes are there to enhance human vision. The term is widely used especially in the literature on computational medicine for algorithms supporting humans in medical diagnosis and research. Long and Ehrenfeld (2020) proposed such an augmentation scenario impressively for the case of reacting to the Corona pandemic (in a paper published at a time when the general public hasn't realized yet that there was a pandemic), forecasting a coordinated research endeavor to fight the spread of the disease that would have been not possible without strong ML capacities supporting the research teams. Reality proofed their predictions right.

However, in the understanding of augmented intelligence, there still is a sequential process in the division of labor between humans and machines: Machines process large amounts of data, search for patterns, and make predictions, but basically support humans, who drive the process, and execute the results of the AI. Our understanding of hybrid intelligences goes further, regarding human and machine intelligence as two elements of a common loop. In doing so, we follow the definition by Dellermann et al. (2019), as presented above, or Zheng et al. (2017), who describe a "human-in-the-loop hybrid-augmented intelligence" system, where humans are always part of the system. In this system, humans first influence the outcome (of a machine intelligence) in such a way that they provide further judgment if a low confident result is given by the algorithm. But the collaboration goes further. The idea is to "realize a close coupling between the analysis-response advanced cognitive mechanisms in fuzzy and uncertain problems and the intelligent systems of a machine" (Zheng et al., 2017: 154). Hence, human and machine intelligence adapt to and collaborate with each other, forming a two-way information exchange and control (a similar understanding has been outlined by Pan (2016) in his conceptualization of an "Artificial Intelligence 2.0"). This is why we prefer to use the term *hybrid* (and not augmented) intelligence.[1]

A good illustration of this collaboration between human and machine intelligences provides AlphaGo, a Go-playing computer developed by DeepMind Technologies (a firm belonging to Alphabet Inc., the mother company of Google). Commonly seen as a breakthrough in machine intelligence, AlphaGo defeated the

---

[1] Zheng et al. (2017) also describe a second concept of human and machine collaboration: "cognitive computing-based hybrid-augmented intelligence." While out of the scope of this chapter, it is worth mentioning. Cognitive computing-based hybrid-augmented intelligence refers to a machine that "mimics the function of the human brain and improves computer's capabilities of perception, reasoning, and decision-making. In that sense, [it] is a new framework of computing with the goal of more accurate models of how the human brain/mind senses, reasons, and responds to stimulus, especially how to build causal models, intuitive reasoning models, and associative memories in an intelligent system" (Zheng et al., 2017: 154).

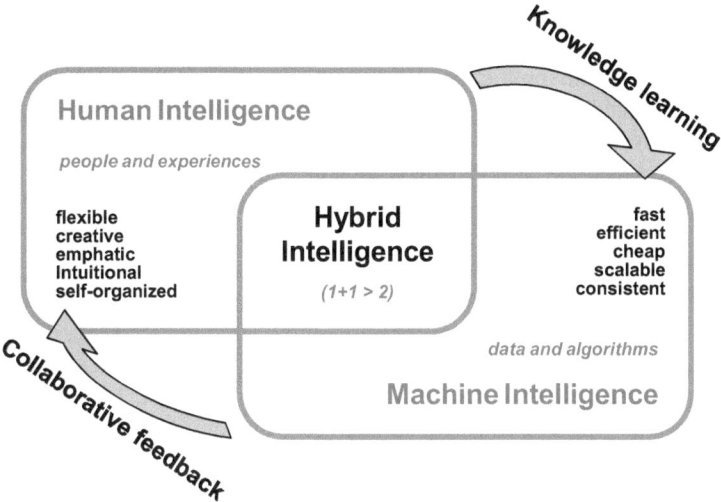

Fig. 1 Hybrid intelligence (HI) aims to combine the best of human intelligence and machine intelligence [Source: Building on van der Aalst (2021) and Zheng et al. (2017)]

best-ranked Go player Ke Jie in 2017. The more powerful AlphaGo Zero learned by just playing games against itself but was able to defeat any human player by the end of 2017. However, this has not been the end of the story (van der Aalst, 2021). The interplay between human intelligence and machine intelligence led to new insights. AlphaGo showed human players new strategies for playing Go, as some of the world's leading Go players acknowledged [as recorded in Baker and Hui (2017)]. Shi Yue said "AlphaGo's game transformed the industry of Go and its players. The way AlphaGo showed its level was far above our expectations and brought many new elements to the game." Zhou Ruiyang said "I believe players more or less have all been affected by Professor Alpha. AlphaGo's play makes us feel freer and no move is impossible to play anymore. Now everyone is trying to play in a style that has not been tried before." At the same time, the new strategies explored by the human players inform the machine algorithm. Humans can learn from machines, and machines from humans: "We look forward with great excitement to AlphaGo and human professionals striving together to discover the true nature of Go," Baker and Hui (2017) conclude a review of the innovations to the gameplay of Go, resulting from the collaboration of human players and the AlphaGo machine.

Hybrid intelligence aims to combine the best of both worlds, as illustrated in Fig. 1. The recent developments in AI and ML have extended the reach of software and hardware automation (robots). Once a robot is able to perform a repetitive task at a similar level of quality, it is often also more reliable and cost-effective. However, humans still have unique capabilities. For example, we have the ability to transfer experiences from one problem domain to another. As van der Aalst et al. (2021) argue, AI/ML cannot deal with disruptions. The Corona pandemic or events of severe weather like the flooding in Germany in July 2021 have shown that when there is a sudden dramatic change, predictive models fail, no matter how much data was there before. Especially at the beginning of the Corona pandemic, the established algorithms predicting demand in supply chains failed because of the unforeseen demand for certain products (e.g., pasta and toilet paper) combined with simultaneous restrictions for travel, work, and business. In such a situation, machine intelligence needs to be complemented by human intelligence.

But also in non-catastrophic events, humans need to remain in the loop. The idea of hybrid intelligence is not just to use humans when machine intelligence fails due to disruptions. The allocation of machine intelligence in decision-making processes often leads to more efficient, but sometimes also to unreflective or non-transparent, solutions with unintended biases. This, in turn, leads to a rejection of the AI contribution (acceptance) and thus hinders the exploitation of its potentials. Consider situations that need empathy, creativity, or ethics (van der Aalst, 2020). Decisions in these situations will also demand human contributions and cannot entirely be executed by a machine. Machine intelligence and human intelligence will complement each other. Understanding these factors as well as the mechanisms of interaction between humans and machine intelligence is a domain that opens a wide demand for further research. We will explore these dynamics in larger detail in the following section.

3 New Rules for Task Allocation: Division of Labor Revisited

The rise of hybrid intelligence asks us to reconsider one of the most fundamental of all economic and ergonomic questions: the division of labor and task allocation in an organization and individual work systems. While the development of machine intelligence is a field of computer science (decision routines and data structures) and research on corresponding technical applications of AI is primarily located in the engineering sciences, the implementation of hybrid intelligence is an economic (management) phenomenon (Bailey & Barley, 2020; von Krogh, 2018). It asks the question how to efficiently design decision-making in an organization.

Since the days of Frederick Taylor and Henry Ford, the idea of the ideal human-machine task division has evolved considerably from an industrial engineering and ergonomic perspective. Machine intelligence has the potential to be more than a tool, as it can also take on the role of a work partner or even a supervisor, as suggested in the debate of algorithmic management (Lee et al., 2015). In a work system, humans and AI need not oppose each other, but can complement each other as a team. Still, today, humans are only used for monitoring systems automated by machine intelligence. These humans are either under-challenged or fatigued, which significantly prompts errors. Other humans, who already are heavily burdened by their own subtasks, get overwhelmed by the need to make additional decisions as to when AI support should be utilized. Hence, to effectively support and relieve humans, machine intelligence should therefore work largely independently and recognize when support is necessary and desired. Furthermore, a dynamic division of tasks between humans and machines could adapt to varying situations, tasks, and user states, avoiding states of cognitive overload and underload. As a basis for such adaptive support, data providing information about the states of the individual components of a work system, like the involved human(s), equipment, the environment, as well as task and organizational goals, are needed and can be provided in the future in the form of digital shadows.

When the extent of decision support by machine intelligence is reaching intensity levels that seemed impossible in the past, research is needed how tasks can be allocated in the continuum between machine intelligence and human intelligence. Prior research in this domain rather described the challenge based on a few case studies (e.g., Iansiti & Lakhani, 2020; De Cremer, 2020) or exploratory surveys (Berditchevskaia & Baeck, 2020) and rather focused on the practical implementation of decision processes with machine intelligence, but neither examine their organizational impact nor do they follow the understanding of a hybrid intelligence, as discussed before.

We propose to structure such a research endeavor into two dimensions:

1. What is the (optimal) degree of integration of machine intelligence into organizational decision processes, and what are the tasks remaining for humans and the tasks where a human-machine collaboration is the preferred solution?
2. What is the quality of decisions made by the use of machine intelligence—not just when compared to the factual quality of the decision for a given task (if benchmarked against human decisions) but also when taking factors of organizational acceptance and adoption of the machine decision into account?

3.1 Degree of Machine Intelligence Integration into Organizational Decision Processes

To analyze the degree to which machine intelligence is involved in organizational decision-making, the established logic of the automation pyramid in engineering provides a good framework (Endsley, 1987). Consider the different cases shown in Fig. 2. The two extremes are the established situations of human and machine intelligence. But as the picture shows, there is a scope of hybrid situations [(b) to (d) in Fig. 2]. Here, to varying degrees of intensity, human and machine intelligence interact, each with particular strengths (and weaknesses) and major differences in capabilities and behaviors, in ways that did not exist in earlier human-human interactions (Berditchevskaia & Baeck, 2020; Groensund & Aanestad, 2020). In a narrow understanding of our definition of hybrid intelligence, only Case (d) addresses the intended collaboration between human and machine intelligence; Cases (b) and (c) are rather situations of "augmented intelligence." However, the

	Decision preparation	Decision	Decision scenario and assessment
Human Intelligence	**Human**		**(a) Human makes the decision alone.** • Ability to judge and interpret ("common sense") also with regard to ethical standards; identification of objectives; creativity, flexibility; emotional intelligence • Possibly technically inadequate, slow; characterized by human bias in decision-making.
Hybrid intelligence	*AI & ML (machine)*	*Human*	**(b) Information is prepared by machine intelligence, but the actual decisions are made by humans ("augmented intelligence")** • Human judgment as a controlling factor when using ability for, e.g., pattern recognition and speed of AI&ML • Acceptance of machine prediction / prescription harmed by interpretation, norm violations, distribution effects, non-transparency
	Human	*AI & ML (machine)*	**(c) Humans provide machines with information, but autonomous decision-making is done by machine intelligence** • Initiation by humans (control of the process), by, e.g. providing pre-selected or pre-labeled data • Solution may contain human bias; not based on sufficient data
	Human-AI (human-machine) Collaboration		**(d) Machine and human intelligence are integrated and collaborate during the decision process and task execution** • Iterative learning between human and machine based on machine-enabled pattern recognition and simulation, as well as human feedback to the machine. • Machine decision patterns shape human decision strategies in the long term
Machine Intelligence	*AI & ML (machine)*		**(e) Decision is fully delegated to an autonomous machine** • Speed, scalability, processing of large amounts of data. Objectivity (i.e. ex ante free from human bias). • Not always technically feasible; erroneous, especially biased decisions with distorted or incomplete data.

Fig. 2 Different situations of combining human and machine intelligence

borders between these areas are fuzzy and constantly moving, as we will illustrate with a simple example at the end of this chapter.

All situations of hybrid intelligence have immediate consequences for the behavior of individuals and thus for the resulting (quality of the) decisions and their implementation. In the longer term, they will also result in indirect effects, when people's experiences with machine intelligence influence their subsequent behavior in other situations (e.g., always expecting that there is a machine intelligence at hand to support a human task). Also, undesirable path dependencies may arise, such as a loss of knowledge or skills (Lebovitz et al., 2022), as experienced by the use of GPS-based navigation systems, which deterred the ability of many humans to navigate without machine support.

Hence, a critical question is when the potential benefits of allocating decision-making tasks to machine intelligence (increasing the efficiency and effectiveness of the decision-making process) are (over)compensated by new costs and challenges. These costs include both the efforts for developing and implementing the algorithms and the cost of adapting an organizational design to the new situation. Also, indirect costs in the form of negative effects on human behavior must be considered, e.g., costs resulting from acceptance problems. Acceptance here addresses both the individual level, i.e., humans who must share decision power with machine intelligence and collaborate with it, and the societal level of acceptance by stakeholder groups such as trade associations, unions, or regulatory institutions.

3.2   Consequences for Decision Quality

For certain, well-defined decision situations and tasks, machine intelligence provides without doubt better results, i.e., adds real value (without obvious violation of norms and other constraints). However, also in these situations, a remaining challenge is often the black-box nature of the solution (Shrestha et al., 2019). In computer science, approaches are therefore being developed to make AI more comprehensible (Rai, 2020), so that people are more likely to accept and implement the solution provided by the machine intelligence (when completely autonomous task performance is not possible/desirable). Scenarios of using hybrid intelligence are obvious in these decision-making situations.

In other situations, however, it is not certain whether machine intelligence can provide a suitable and better solution. This may be because (1) relevant norms to the decision are not observed by the machine and/or (2) the technical solution is "flawed," because the underlying data basis is insufficient or the modeling has not adequately captured the problem or cannot capture it due to unknown causal relationships. An example of such flawed decisions can be found in recruitment. When past career paths and performance patterns are used as the basis for future hiring, women tend to be left out of the equation (Cowgill & Tucker, 2020). This results in a conflict with the social norm of increasing diversity. The reasons behind these flawed decisions can be insufficient amounts of data or discriminatory patterns contained therein, but also an ill-defined notion of recruitment performance. However, once such a problem has been understood, humans together with machine intelligence can improve automating these decisions in the mid-term.

We believe that this situation also reflects the reality in most manufacturing companies today (Agrawal et al., 2019; Raj & Seamans, 2019). Machine intelligence is used but requires collaboration with human decision-makers to result in an optimal solution. Hence, an important question is how humans could check the quality of prescriptions provided by a machine, considering a potential violation of norms or possible "errors," before implementing the solution in a corrected manner, a procedure that Groensund and Aanestad (2020) called "augmenting the algorithm." As we will argue in the next section, real-time simulation models enabled by digital twins and shadows allow exactly such an ex ante validation. At the same time, structuring a machine intelligence solely according to human thought patterns (or those that humans can understand) is not sufficient either, as it may model the problem task inadequately or follow violations of norms by human decision-makers. This is exactly where the vision of a hybrid intelligence comes into place. Once the issues outlined before are recognized and understood, either an autonomous decision process by machine intelligence could be improved, or the decision could be structured in such a way that humans stay in the loop, taking social norms or intended consequences into account. Equally, however, humans also improve their own decision-making processes, when, for example, a machine intelligence suggests previously unknown initial solutions or uncovers distorted decision-making patterns of humans in the past. The loop is closing.

4   Hybrid Intelligence in Next Generation Manufacturing

While we believe that hybrid intelligence will strongly influence all kinds of decisions and task execution in an organization, we want to demonstrate such a scenario for Next Generation Manufacturing, as central to this book. As introduced in chapter "How Digital Shadows, New Forms of Human-Machine Collaboration, and Data-Driven Business Models Are Driving the Future of Industry 4.0", the context of this work is the interdisciplinary research cluster *Internet of Production (IoP)* at *RWTH Aachen University* (iop.rwth-aachen.de), enabling a new level of cross-domain collaboration along the entire product life cycles from engineering over operations toward the usage stage (Brecher et al., 2016). The IoP pursues a vision called the World Wide Lab (WWL), in which processes, factories, entire companies, and the managers and workers constituting these organizations can learn from each other by sharing experiences and knowledge (Brauner et al., 2022). Corresponding to the relationship of the Internet and the World Wide Web (WWW), the WWL aims to be a network of multisite labs in which models and data from experiments, manufacturing, and usage are made accessible across company borders to gain additional knowledge. A main driver of the WWL is *digital shadows*, i.e., purpose-driven, aggregated, multi-perspective, and persistent data sets from production, development, or usage (Liebenberg & Jarke, 2020). Digital shadows are a specification of the broader idea of digital twins (for more details, refer to chapter "How Digital Shadows, New Forms of Human-Machine Collaboration, and Data-Driven Business Models Are Driving the Future of Industry 4.0"). The cross-domain exchange of digital shadows in the form of *data spaces* can make data more valuable, opening the present data silos in different companies—a core enabler of better machine intelligence.

In our understanding of the Internet of Production, digital shadows are the "units of data" shared among organizations. They connect data, products, and industrial assets within and across organizations and are the foundations for data-driven planning and decisions within an organization (factory) and in-between organization (supply chains, value chains) by using real-time and historical data to simulate predicted futures. In this loop, hybrid intelligence plays a central role. Figure 3

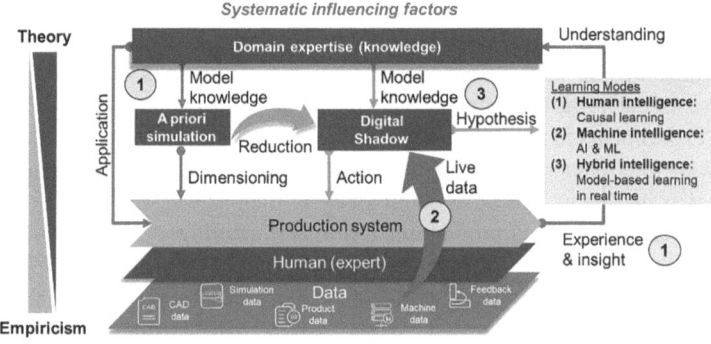

Fig. 3 Three models of learning in manufacturing: (1) causal learning, (2) machine learning, and (3) model-based learning based on digital shadows [building on Brecher et al. (2017)]

outlines such a *hybrid decision-making* combining machine intelligence and human expertise in a collaborative form. The figure shows three different modes of learning (understood here generically as any kind of decision-making in an existing manufacturing system to improve the system's operational efficiency, to cope with disturbance, or to increase the system's potential for strategic differentiation).

1. *Human intelligence: Causal learning* is the established way we learn. Building on *domain knowledge* acquired either by experience (learning on the job) or by formal education, humans have a unique capability to understand a complex system and utilize or improve it by trial-and-error learning. The experience curve effect is based on this learning mode, as are practices like Lean Six Sigma. Informed by their domain expertise, a team at a production station defines a problem area (an *application*), sets up *assumptions* (hypotheses) how to achieve an improvement, tests the assumptions via *experiments* to gain *insights*, and then either implements the solution (if the experiment was successful) or redefines the assumptions and conducts a new experiment. The development of the hypotheses is based on *theory*, often captured in models of the systematic influencing factors of the production system (like fluid or thermal dynamics) and uses the real production system as the test bed for empirical validation (*empiricism*). Such a causal learning process can be very powerful, but it is often slow and prone to the assumptions human draw and the hypotheses they set up.

   Conventional (digital) simulation models also belong to this learning mode. An *a priori simulation* uses ("theoretical") model knowledge to simulate (an extract of) the production system, so that specific behaviors (assumptions, scenarios) can be tested. These digital models can be used to reason about reality and answer what-if questions. However, digital simulation models are a reflection of reality that is created manually and functions in an offline manner, i.e., the model does not change when reality changes (van der Aalst et al., 2021). Hence, conventional simulation models are outdated when the production system goes into operations, as there are numerous *stochastic factors* influencing the system behavior, like temperature conditions, material characteristics, or the mood of the humans involved. All of these factors lead to a continuous change of the system (like an abrasion of a component, minimal modifications of a material, etc.). Conventionally, these changes are not captured in the simulation model, which is why experiments in the real system are required.

2. *Machine intelligence: Artificial intelligence and machine learning* came up as the new way to learn. Machine intelligence is data-driven and learns from data without explicitly being programmed. In manufacturing, cyber-physical systems provide these data in real time (in the form of digital twins and shadows) and store it in repositories (data spaces) where algorithms can find insights and interrelations between different data sets. Today's usage of machine learning for many tasks that before could only be done by humans can be attributed to progress in deep learning techniques, where *artificial neural networks (ANNs)* having multiple layers progressively extract higher-level features from the raw input (van der Aalst, 2021). For example, we can train an ANN to distinguish between pictures from a vision control system that show work pieces with adequate and others with insufficient quality. While training, the ANN updates the weights in the internal representation until the number of incorrectly classified pictures is minimized. Then the trained ANN is used to classify test data, i.e., unseen pictures of good and bad pieces that need to be classified correctly. Given enough training data, such an ANN may perform amazingly well in automating quality control, although it was never programmed to do so and has no explicit knowledge of what makes a good and a bad work piece.

   Beyond such automation scenarios enabled by machine intelligence, also higher-level learning can take place. When the quality data (from the automated vision control system) is matched with data from other workstations of this production system, algorithms can find patterns between two system elements, identifying also states in one production step that causes later whether a work piece is labeled as good or bad. This ability of finding patterns in huge data sets led some people to say that the future of learning in manufacturing is only pattern recognition in huge data spaces—no human input and no domain knowledge required. However, we believe—and were confirmed by the results of our Delphi study—that such a pure machine intelligence scenario is unlikely to cope with the complexity of a real production system.

3. *Hybrid intelligence: Model-based learning in real time* is our proposed scenario for learning in Next Generation Manufacturing. Without doubt, machine intelligence can perform repetitive operational tasks more efficiently than humans can. Machine learning algorithms also have an unmatched capability of finding patterns in large data sets. We propose that these insights generated by *machine intelligence* serve as a highly educated "hunch" for humans, who combine it with their domain expertise on a higher level. An important component of this approach is the availability of *digital shadows* as virtual, real-time digital counterparts of something that exists in the physical world (e.g., a production system, workstation, or work piece). The digital counterpart should help to make decisions in a better way, by not providing the real-time data from which a machine

intelligence can generate its insights, but also the test bed where ideas for improvement and optimization can be validated virtually.

Consider the quality example from Scenario (2). Let us assume that an algorithm provided an insight in the form of a prediction on the causes of a quality issue: "When the temperature in Station A dropped below a specific threshold, later quality errors occurred in Station E." With the availability of a digital shadow, and different to conventional digital simulation models as in Scenario (1), the model behind the digital shadow is automatically derived and changes when reality changes. The digital shadow can now be used to reason about reality and answer what-if questions. Hence, assumptions on how to improve the quality issue in our example can be tested virtually in the simulation models embedded in the digital shadow. This connects Scenarios (1) and (2). Based on their *intuition* and *domain expertise*, human decision-makers could make conclusions on how to improve the quality of the system, e.g., different approaches to control the temperature in Station E in a more stable way or approaches to counterbalance the temperature effect on to work pieces in later word stations. These assumptions about how to improve the system's quality, provided by *human intelligence* but augmented by insights generated by *machine intelligence*, could now be validated in the virtual shadow. The virtual experimentation allows testing of many more alternative scenarios for improvement. A machine intelligence could support this experimentation, e.g., by proposing different scenarios and predicting their outcomes.

In a further state, an *automated real-time feedback loop* can be established. The insights produced by the digital shadow could then either automatically trigger changes in the production system or become implemented manually by humans after interpreting the results (van der Aalst et al., 2021). Results of the digital shadow directly affect reality. For operational situations, autonomous learning and optimization is likely. For example, when the simulation model predicts a delay, the production process could be reconfigured automatically (similar to the re-routing algorithm in a navigation system when it is informed about an incident on the originally planned route). For more complex learning scenarios, like restructuring the manufacturing system or coping with disruptions, the advanced simulation model embedded in a digital shadow allows human decision-makers to evaluate all possible decisions in the virtual world without causing harm, waste, and costs in the real (physical) system. With cheaper and richer experimentation, the likelihood of finding a better solution increases.

We have to stress that this scenario is a picture of the future yet, especially when we apply it on the level of a larger system. In our research in the Internet of Production cluster at RWTH Aachen, our colleagues were able to demonstrate this approach on the component and work station level (Brecher et al., 2019; Xi et al., 2021). *Process mining* can serve as a concrete technology to facilitate the development of such a virtual shadow/twin of an entire system (van der Aalst, 2016). Using process discovery, so-called control-flow models can be derived. Aligning these models with event data, it is possible to add different perspectives (time, costs, resources, decisions, etc.). The resulting elaborate model can be simulated. Using process mining, it is relatively easy to create a digital shadow in terms of a frequently updated virtual replica of a physical object. However, it still is rather difficult to create a model that behaves like a real system, where multiple processes interact and compete for resources concurrently. To fulfill the vision of a digital shadow that automatically takes action, action-oriented process mining provides initial ideas (e.g., the *Celonis Execution Management System* can trigger corrective workflows using the *Integromat* integration platform). But despite these initial capabilities of process mining, it is fair to say that this scenario of hybrid intelligence is more a vision than a reality. We need to keep humans in the loop (Abdel-Karim et al., 2020) to cope with the complexities of an entire production system. This is why we regard hybrid intelligence as the most likely scenario for decision-making the factory of the future.

5 A Simple Application of Hybrid Intelligence in Publishing

We want to close this chapter by a simple use case of hybrid intelligence. When writing and producing this book, we recruited an AI as a member of our author team, tasking it with creating all abstracts of this book's chapters and writing the book's preface. This worked amazingly simple, providing us a real glimpse into a future where machines and humans collaborate intuitively.

The AI we used is a transformer-based language model. While quantitative data prevails in a production context, much knowledge is shared through natural language. By talking to a colleague, listening to a lecture, or reading a book, understanding language grants us access to a plethora of knowledge. Today, AI has reached a good level of language understanding, so that we can use such technologies to further share and create knowledge. This makes language models an especially interesting form of AI to use in knowledge-intensive work (Bouschery et al., 2022).

Transformer-based language models are a special kind of AI used for natural language processing (NLP), which Liddy (2018: 3346) defines as a range of "computational techniques for analyzing and representing naturally occurring texts at one or more levels of linguistic analysis for the purpose of achieving human-like language processing for a range of tasks or applications." In general, natural language processing is not new to firms. It has been used, for example, in text analysis (text mining), like generating insights from maintenance or service reports. Prior models have typically been very task specific. Also in this field, a great deal of progress stems from advances in ANNs. Newer NLP technologies show the potential to take on multiple knowledge-related tasks and cannot just analyze existing text but also generate new one. A core example of these advanced NLP models is generative or transformer-based models. At its core, language modeling is the process of predicting the next word in a sequence based on its preceding characters or words. This field has seen continued progress over the past decades with a trend toward larger and more complex models rapidly increasing the models' capabilities—from the mere suggestion of related words to state-of-the-art models that can produce full newspaper articles indistinguishable from human-written text (Brown et al., 2020).

New transformer-based models contain attention mechanisms that allow for parallelization during the processing of inputs and thereby eliminate some of the main performance issues of recurrence-based models, leading to significantly faster models (Vaswani et al., 2017). Another big advantage of these types of models is their ability to take the context of the processed words into consideration, which allows for a far more nuanced understanding of related words and concepts within a given text and subsequently more complex applications. Today, most of the state-of-the-art language models are based on this transformer architecture and rely on large data sets only for pre-training purposes. Examples include Google's BERT model or OpenAI's line of Generative Pre-trained Transformers (GPT). The number of parameters used to generate the models' output has increased significantly over the last few years. For example, the original BERT model (Devlin et al., 2018) uses 340 million parameters in its largest instance. This pales in comparison to OpenAI's latest model, GPT-3. In just 3 years, the model size of the GPT line has grown by nearly 1600% from 110 million parameters in the original model over 1.5 billion parameters in its second iteration (Radford et al., 2019) to 175 billion parameters in GPT-3 (Brown et al., 2020). The next version is expected to have 100 trillion parameters. Because transformer-based language models' capabilities significantly improve with model size, the rapid increase in model sizes has dramatically increased the usefulness and applicability of such transformer-based language models.

While it is very cost-intensive to build and train large transformer-based language models in the first place, many of these models have been open-sourced and can be accessed very easily through web services, making them accessible to a broader audience. Also, commercial applications like GPT-3 are available in cloud-based applications via a standard Internet browser. Another big advantage of transformer-based language models is that users can generally interact with them simply through natural language. Companies like OpenAI provide access to their models through not only application programming interfaces (APIs) but also graphical user interfaces (GUIs), which significantly lower the barriers to entry.

For this book, we utilized OpenAi's GPT-3 [we refer to Bouschery et al. (2022) for a more detailed description of our approach]. To interact with the model, users have to provide some initial text input. This could either be a question, the beginning of a story that should be completed, some text that should be summarized, bullet points to turn into written text, etc. Based on this initial input and its knowledge learned during training, GPT-3 then generates a text that best fits the provided prompt by predicting the next word in the sequence based on the previous words in the prompt. GPT-3 is a so-called few-shot learner, which means that users are advised to provide the model with a few examples to show what kind of output they expect from the model. The initial prompt is therefore the main way of steering the model toward a desired output—*a perfect illustration of a hybrid intelligence.*

We hired the GPT-3 to become a member of our publishing team for a typical knowledge processing task: knowledge extraction (De Silva et al., 2018), i.e., making existing knowledge usable by extracting knowledge that might be coded explicitly or implicitly in a given knowledge base. Normally, extracting knowledge is rather labor-extensive and not easily scalable. However, transformer-based language models provide the opportunity to automate parts of such processes. The knowledge task we asked the GPT-3 to do was *text summarization*. All abstracts of this book (including the one of this chapter) were generated by the GPT-3. Extracting knowledge from text requires the identification, selection, and processing of large amounts of textual data, which can be very costly and time-consuming as well as overwhelming for humans. This might lead to unwanted consequences like bad decision-making, as humans tend to make simplifying assumptions in order to be able to process such large volumes of information (Simon, 1955). Transformer-based language models provide an adequate solution to such problems through their ability to summarize text efficiently while conveying the most important aspects and meaning of the original text. They are especially well suited for such tasks because of their context awareness and the ability to pay special attention to the most relevant pieces of information. This gives them a great advantage over earlier text summarization algorithms, which struggled to understand sentiment and meaning in textual data (Fan et al., 2006). We invite our readers to evaluate the results by themselves.

Furthermore, users can prompt state-of-the-art language models like GPT-3 to summarize a text in a very specific way. In addition to the abstracts provided in each chapter of this book, we also tasked GPT-3 to summarize the entire book as short as possible to test an extreme case of summarization. This is the result [the differences result from varying settings of the model, see Bouschery et al. (2022), for more details].

Prompt: "GPT-3, summarize this book in a few sentences."

- The text discusses different projecting made by experts about the changes that will occur in production by 2030. These include the trend of increasing digital goods, the use of hybrid intelligence to supplement human decision-making processes, the establishment of full digital twins of each production worker and all of their operations, and the decentralization of supply chains.
- The text discusses various trends that are projected to occur in the manufacturing industry by 2030. These trends include the use of subscription models for production machines, the use of collaborative robots, the use of digital twins, and the decentralization of supply chains. Each of these trends is expected to have a significant impact on firms.
- The Delphi study found that in 2030, subscription models for production machines will be the new industry standard, that environmental sustainability will play a bigger role, and that AI-based decision-making systems will not be able to significantly increase production resilience.

Prompt: "GPT-3, summarize this book in one sentence."

- The text discusses various projections for the year 2030 and their potential impacts on different aspects of society and the economy.

These examples show the ability of transformer-based language models to capture the essence of a text and reproduce a coherent, yet substantially summarized, version of the original text. We did not provide any further context and examples or used fine-tuning to produce the response. While this task might seem rather trivial to a human, it is rather complex. First, the model had to understand that there is a difference between the provided article and the task description. Specifically, the task descriptions are, in fact, exactly that, a description of what the model should do, and not, e.g., the beginning of a novel that should be continued. Then, it had to analyze, understand, and then generate a summary of the said abstract that was factually, semantically, and grammatically correct. All, without having specifically been trained to perform this task. Noteworthy is also that the model did not just shorten the provided text, but that it summarized the text in its own words. However, when looking closely at the generated texts, we instantly find expressions which we would write differently, where there would be a dedicated technical term to describe the subject more precisely for an expert audience, or where we also would emphasize an aspect we believe being most interesting for our target audience of academic peers (who the algorithm does not know at all).

Hence, we propose that transformer-based language models will specifically support knowledge-based practices in the form of a hybrid intelligence. Their ability to interact with different knowledge sources, to learn from them, and to transform knowledge allow these models to act as a knowledge broker that facilitates the sharing of knowledge between different stakeholders while also fostering the creation of new knowledge (Waardenburg et al., 2022). Human teams can employ these language models to access existing knowledge. Models that have been trained on large text corpora from the Internet have knowledge on a wide range of topics, which opens up the opportunity for teams to integrate knowledge that might lay outside their area of expertise. Given a prompt by a human, the AI can help to establish connections between concepts and ideas that might otherwise not have been obvious. Few-shot learning capabilities then allow for an easier interaction between the humans in a hybrid team and the AI. Humans have to provide a limited number of exemplary responses to a given task, so that the language model can generate a first adequate output. Humans then evaluate this output, indicating to the algorithm, for example, parts of the output they find especially interesting. The algorithm will then produce a next output, based on this feedback. In the true understanding of a hybrid intelligence, machines and humans are building upon each other's input and output.

In such a scenario, teams can integrate the AI in their existing processes, as if it would be a new colleague. The combination of domain expertise by human team members and knowledge provided by the AI provides the opportunity to greatly improve productivity of knowledge-based practices and produce outcomes that would not have been possible with just the skillset of one of the actors. Orchestrating and building such hybrid teams becomes a new important managerial task, and understanding when and how to allocate tasks to a machine intelligence (and which one) will be a key success factor of organizations in the future. Managers have to consider the distinct characteristics of human and non-human actors. While humans will play a major role in providing context, steering language models toward desired results, and embedding AI output in the larger picture, machine intelligence can speed up many tasks that require the handling of large amounts of text (or other data), understand patterns in data invisible to humans, and make connections between knowledge bases that might not be readily available to human team members.

While for more complex tasks like steering a production system such a scenario of hybrid intelligence is still not existing, the way of development seems clear. We hope that this chapter, but also the analysis of our Delphi study in the entire book, provides the reader plenty of ideas and food for thought about the future of industrial production and the elements of Next Generation Manufacturing.

Acknowledgment Funded by the Deutsche Forschungsgemeinschaft (DFG, German Research Foundation) under Germany's Excellence Strategy—EXC-2023 Internet of Production—390621612.

References

Abdel-Karim, B. M., Pfeuffer, N., Rohde, G., & Hinz, O. (2020). How and what can humans learn from being in the loop? Invoking contradiction learning as a measure to make humans smarter. *Künstliche Intelligenz, 34*(2), 199–207. https://doi.org/ghqvr8

Agrawal, A., Gans, J., & Goldfarb, A. (2019). Artificial intelligence: The ambiguous labor market impact of automating prediction. *Journal of Economic Perspectives, 33*(2), 31–50. https://doi.org/ggnh5t

Bailey, D., & Barley, S. R. (2020). Beyond design and use: How scholars should study intelligent technologies. *Information and Organization, 30*(2), 100286. https://doi.org/hm62

Baker, L., & Hui, F. (2017, April). *Innovations of AlphaGo*. Research blog by Deepmind. https://deepmind.com/blog/article/innovations-alphago

Berditchevskaia, A., & Baeck, P. (2020). *The future of minds and machines: How AI can enhance collective intelligence*. Nesta Report.

Bouschery, S., Blazevic, V., & Piller, F. (2022). *Artificial intelligence as an actor in hybrid innovation teams: An assessment of the GPT-3 language model*. Forthcoming as a Catalyst Paper in the Journal of Product Innovation Management.

Brauner, P., Dalibor, M., Jarke, M., Kunze, I., Koren, I., Lakemeyer, G., ... Ziefle, M. (2022). A computer science perspective on digital transformation in production. *ACM Transactions on Internet of Things, 3*(2), 1–32.

Brecher, C., Eckel, H. M., Motschke, T., Fey, M., & Epple, A. (2019). Estimation of the virtual work piece quality by the use of a spindle-integrated process force measurement. *CIRP Annals, 68*(1), 381–384. https://doi.org/hm63

Brecher, C., Özdemir, D., & Weber, A. R. (2016). Integrative production technology: Theory and applications. In C. Brecher & D. Özdemir (Eds.), *Integrative production technology* (pp. 1–17). Springer. https://doi.org/hhn9

Brecher, C., et al. (2017). Learning production systems. In *Proceedings of the 29th AWK Aachener Werkzeugmaschinen-Kolloquium* (pp. 135–161). Apprimus.

Brown, T. B., Mann, B., Ryder, N., Subbiah, M., Kaplan, J., Dhariwal, P., Neelakantan, A., Shyam, P., Sastry, G., & Askell, A. (2020). *Language models are few-shot learners*. ArXiv:2005.14165.

Cowgill, B., & Tucker, C. E. (2020). *Algorithmic fairness and economics*. Columbia Business School Research Paper.

De Cremer, D. (2020). *Leadership by algorithm: Who leads and who follows in the AI era.* Harriman House.

De Silva, M., Howells, J., & Meyer, M. (2018). Innovation intermediaries and collaboration: Knowledge–based practices and internal value creation. *Research Policy, 47*(1), 70–87. https://doi.org/gcshf5

Dellermann, D., Ebel, P., Söllner, M., & Leimeister, J. M. (2019). Hybrid intelligence. *Business and Information Systems Engineering, 61*(5), 637–643. https://doi.org/ggkxz4

Devlin, J., Chang, M.-W., Lee, K., & Toutanova, K. (2018). BERT: Pre-training of deep bidirectional transformers for language understanding. *ArXiv, 1810,* 04805. https://doi.org/hm65

Endsley, M. R. (1987). The application of human factors to the development of expert systems for advanced cockpits. *Proceedings of the Human Factors Society Annual Meeting, 31*(12), 1388–1392. https://doi.org/fzdz4g

Fan, W., Wallace, L., Rich, S., & Zhang, Z. (2006). Tapping the power of text mining. *Communications of the ACM, 49*(9), 76–82. https://doi.org/b7f48c

Groensund, T., & Aanestad, M. (2020). Augmenting the algorithm: Emerging human-in-the-loop work configurations. *Journal of Strategic Information Systems, 29*(2), 101614. https://doi.org/gjjp64

Hirsch-Kreinsen, H., & Ittermann, P. (2021). Digitalization of work processes: A framework for human-oriented work design. In *The palgrave handbook of workplace innovation* (pp. 273–293). Palgrave Macmillan.

Iansiti, M., & Lakhani, K. R. (2020). Putting AI at the firm's core. *Harvard Business Review, 98*(1), 59–67.

Kamar, E. (2016 July). Directions in hybrid intelligence: Complementing AI systems with human intelligence. In *Proceedings of the twenty-fifth international joint conference on artificial intelligence* (pp. 4070–4073).

Lebovitz, S., Lifshitz-Assaf, H., & Levina, N. (2022). To engage or not to engage with AI for critical judgments: How professionals deal with opacity when using AI for medical diagnosis. *Organization Science, 33*(1), 126–148. https://doi.org/gn3jks

Lee, M. K., Kusbit, D., Metsky, E., & Dabbish, L. (2015). Working with machines: The impact of algorithmic and data-driven management on human workers. In *Proceedings of the 33rd annual ACM conference on human factors in computing systems* (pp. 1603–1612).

Liddy, E. D. (2018). Natural language processing for information retrieval. In J. D. McDonald & M. Levine-Clark (Eds.), *Encyclopedia of library and information sciences* (Vol. 5, 4th ed., pp. 3346–3355). CRC Press.

Liebenberg, M., & Jarke, M. (2020). Information systems engineering with digital shadows: Concept and case studies. In S. Dustdar, E. Yu, C. Salinesi, D. Rieu, & V. Pant (Eds.), *Advanced information systems engineering. CAiSE 2020* (Lecture notes in computer science) (Vol. 12127). Springer. https://doi.org/hhph

Long, J. B., & Ehrenfeld, J. M. (2020). The role of augmented intelligence (AI) in detecting and preventing the spread of novel coronavirus. *Journal of Medical Systems, 44*(3), 1–2. https://doi.org/ggp6f3

Mütze-Niewöhner, S., Mayer, C., Harlacher, M., Steireif, N., & Nitsch, V. (2022). Work 4.0: Human-centered work design in the digital age. In W. Frenz (Ed.), *Handbook industry 4.0: Law, technology, society.* Springer.

Pan, Y. (2016). Heading toward artificial intelligence 2.0. *Engineering, 2*(4), 409–413. https://doi.org/gfwwrf

Radford, A., Wu, J., Child, R., Luan, D., Amodei, D., & Sutskever, I. (2019). Language models are unsupervised multitask learners. *Open AI blog, 1*(8), 9.

Rai, A. (2020). Explainable AI: From black box to glass box. *Journal of the Academy of Marketing Science, 48*(1), 137–141. https://doi.org/ggw7h2

Raj, M., & Seamans, R. (2019). Primer on artificial intelligence and robotics. *Journal of Organization Design, 8*(1), 1–14. https://doi.org/hm67

Shrestha, Y. R., Ben-Menahem, S., & Von Krogh, G. (2019). Organizational decision-making structures in the age of artificial intelligence. *California Management Review, 61*(4), 66–83. https://doi.org/gf48d3

Simon, H. A. (1955). A behavioral model of rational choice. *The Quarterly Journal of Economics, 69*(1), 99–118. https://doi.org/dw3pfg

van der Aalst, W. M. (2016). *Process mining: Data science in action*. Springer.

van der Aalst, W. M. (2020). On the Pareto principle in process mining, task mining, and robotic process automation. In *Proceedings of the 9th international conference on Data Science, Technology and Applications (DATA 2020)* (pp. 5–12). https://doi.org/hm7b

van der Aalst, W. M. (2021). Hybrid Intelligence: To automate or not to automate, that is the question. *International Journal of Information Systems and Project Management, 9*(2), 5–20. https://doi.org/gk92bq

van der Aalst, W. M., Hinz, O., & Weinhardt, C. (2021). Resilient digital twins. *Business and Information Systems Engineering, 63*(6), 615–619. https://doi.org/gmv8sh

Vaswani, A., Shazeer, N., Parmar, N., Uszkoreit, J., Jones, L., Gomez, A. N., Kaiser, Ł., & Polosukhin, I. (2017). Attention is all you need. In *Proceedings of the 31st conference on neural information processing systems*. NIPS.

Von Krogh, G. (2018). Artificial intelligence in organizations: New opportunities for phenomenon-based theorizing. *Academy of Management Discoveries, 4*(4), 404–409. https://doi.org/gfztxx

Waardenburg, L., Huysman, M., & Sergeeva, A. V. (2022). In the land of the blind, the one-eyed man is king: Knowledge brokerage in the age of learning algorithms. *Organization Science, 33*(1), 59–82. https://doi.org/gntnhp

Xi, T., Benincá, I. M., Kehne, S., Fey, M., & Brecher, C. (2021). Tool wear monitoring in roughing and finishing processes based on machine internal data. *International Journal of Advanced Manufacturing Technology, 113*(11), 3543–3554. https://doi.org/gndbwx

Zheng, N. N., Liu, Z. Y., Ren, P. J., Ma, Y. Q., Chen, S. T., Yu, S. Y., & Wang, F. Y. (2017). Hybrid-augmented intelligence: Collaboration and cognition. *Frontiers of Information Technology and Electronic Engineering, 18*(2), 153–179. https://doi.org/gg6r35